Precision Landmark Location for Machine Vision and Photogrammetry

José A. Gutierrez
Brian S.R. Armstrong

Precision Landmark Location for Machine Vision and Photogrammetry

Finding and Achieving the Maximum Possible Accuracy

Springer

José A. Gutierrez, PhD
Emerson Corporation
Corporate Technology
St. Louis
Missouri
USA

Brian S.R. Armstrong, PhD, P.E.
Department of Electrical Engineering
and Computer Science
University of Wisconsin – Milwaukee
Milwaukee
Wisconsin
USA

British Library Cataloguing in Publication Data
Gutierrez, Jose A.
Precision landmark location for machine vision and photogrammetry : finding and achieving the maximum possible accuracy
1. Computer vision 2. Photogrammetry - Digital techniques
I. Title II. Armstrong, Brian Stewart Randall
006.3'7
ISBN-13: 9781846289125

Library of Congress Control Number: 2007936782

ISBN 978-1-84628-912-5 e-ISBN 978-1-84628-913-2
Printed on acid-free paper

9 8 7 6 5 4 3 2 1

Springer Science+Business Media
springer.com

To Alan Alejandro and Natacha

To Claire, Amanda, and Beatrice

Preface

Precision landmark location in digital images is one of those interesting problems in which industrial practice seems to outstrip the results available in the scholarly literature. Approaching this problem, we remarked that the best-performing commercial close-range photogrammetry systems have specified accuracies of a part per 100,000, which translates to 10–20 millipixels of uncertainty in locating features for measurement. At the same time, articles we identified in the academic literature didn't seem to give any firm answer about when this level of performance is possible or how it can be achieved.

We came to the problem of precision landmark location by a process that must be familiar to many: out of a desire to calibrate an image-metrology test bench using landmarks in well-known locations. It is straightforward to perform the calibration, minimizing errors in a least squares sense; but we were also interested to know what fraction of the residual errors should be attributed to the determination of landmark locations in the images, and whether this error source could be reduced.

To address these questions, we knew we had to go beyond the consideration of binary images. At the limit of sensitivity, it is clear that all of the grayscale information must be used. Likewise, we knew we had to go beyond an idealized model of image formation that considers the pixels to be point-wise samples of the image; since we were looking for changes to the image that arise with landmark displacements of a small fraction of a pixel width, the details of the photosensitive area within each pixel were going to play a role. Finally, rather than focusing on the performance of a particular algorithm, it seemed better to pursue the Cramér–Rao lower bound, and thereby determine an algorithm-independent answer to the question. There was, after all, a sheer curiosity to know whether part-per-100,000 accuracy is really possible for close-range photogrammetry with digital images.

In a convergence of necessity and opportunity, it turns out that considering a richer model of the image formation process is itself instrumental in making the calculation of the uncertainty tractable. The smoothing introduced into the image by diffraction and other sources has been neglected in some past investigations, which have idealized the digital image as point-wise samples from a discontinuous description of the image. Far from being an unwanted complication, the smoothing makes the calculations feasible. If the image were discontinuous, we would have been obliged to represent it with a smoothed approximation in order to calculate the Cramér–Rao lower bound.

With the Cramér–Rao bound in hand, it is possible to determine the gap between the performance of well-known algorithms for landmark location and the theoretical limit, and to devise new algorithms that perform near the limit.

In response to the question of whether part-per-100,000 measurement is possible from digital images, the reader is invited to turn the page and join us in exploring the limits to precision landmark location.

Contents

1 Introduction

Many applications in machine vision and photogrammetry involve taking measurements from images. Examples of image metrology applications include, in medicine: image-guided surgery and multimode imaging; in robotics: calibration, object tracking, and mobile robot navigation; in industrial automation: component alignment, for example for electronic assembly, and reading 2-D bar codes; and in dynamic testing: measurements from high-speed images. In these applications, landmarks are detected and computer algorithms are used to determine their location in the image. When landmarks are located there is, of course, a degree of uncertainty in the measurement. This uncertainty is the subject of this work.

Examples of landmarks used for metrology are shown in Figures 1.1–1.4. In Figure 1.1, we see a standard artificial landmark on the head of the crash test dummy. Quadrature landmarks such as this one are recorded in high-speed images during tests, and by locating the landmarks, measurements of motions can be taken from the images. In another application, landmarks such as those seen in Figure 1.2 are sometimes implanted for medical imaging prior to surgery. Fiducial marks are landmarks used for registration. This and other terms are defined in a glossary provided in Appendix B. Amongst other uses, these fiducial marks in medical images and associated image-based measurement tools can permit the surgeon to

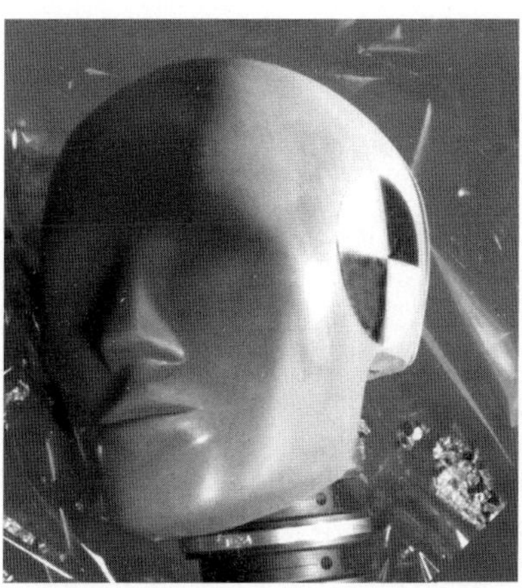

Figure 1.1. Crash test dummies carry *artificial* landmarks to improve the accuracy of motion measurements from images (Photo courtesy of NASA)

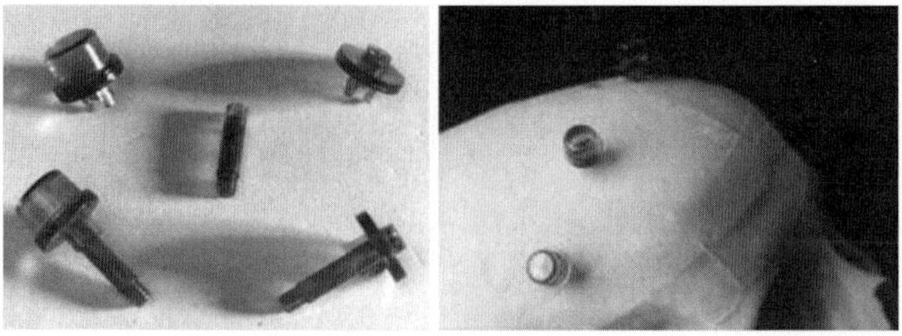

Figure 1.2. Fiducial markers for implanting and precision location in medical images. These markers produce circular patterns in CAT images (Images courtesy of IEEE, from C.R. Mauer *et al.*, "Registration of head volume images using implantable fiducial markers," IEEE Trans. Medical Imaging, Vol. 16, No. 4, pp. 447–462, 1994).

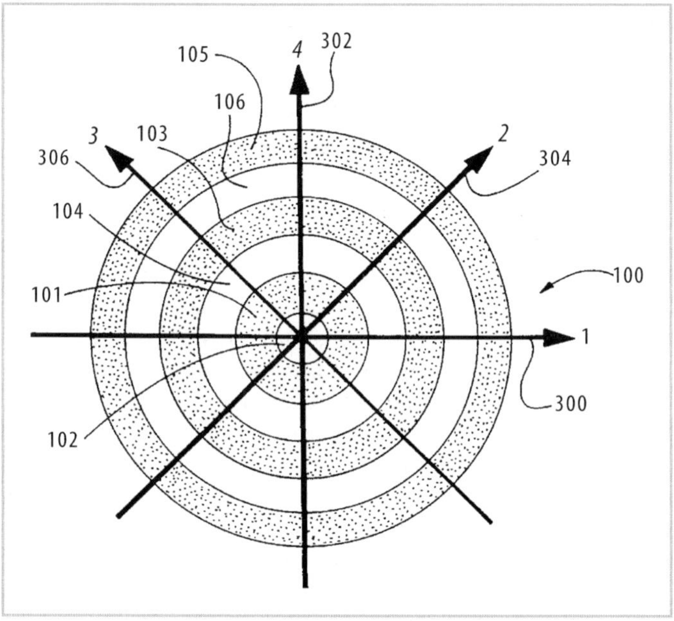

Figure 1.3. Bullseye landmark for robust detection and precision location. This is the landmark used for reading bar-codes on UPS labels. This is the landmark used for reading bar-codes on UPS labels (From U.S. patent 5,515,447, "Method and apparatus for locating an acquisition target in two-dimensional images …").

coordinate the patient on the operating table with 3-D medical images, guiding the way to structures that cannot be seen from the surface. When these tools are used in the operating room, the precision of landmark location plays a role in determining the working volume of the camera-based measurement system, and the flexibility to locate the system out of the surgeon's way. Other examples of landmarks used for industrial metrology or as part of a bar-code reader can be seen in Figures 1.3 and 1.4.

Photogrammetry is the science of spatial measurement from images. In photogrammetry, artificial landmarks can be used to mark and locate features for measurement. Machine vision finds application in many industries where landmarks are used for alignment and inspection. Plate 1 (see Color Section, page 145)

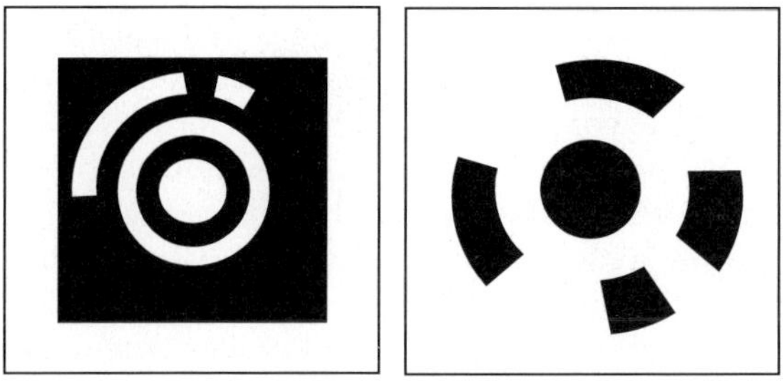

Figure 1.4. Artificial landmarks for metrology (Images courtesy of G. Ganci, H. Handley, "Automation in videogrammetry," Melbourne, FL: Geodetic Services Inc.)

shows artificial landmarks used for alignment or measurement steps in the printing and electronic industries.

The digital images considered in this work are broken into pixels, or picture elements, which are the individual samples of the scene, as seen in Figure 1.5. The challenge of landmark location is to accurately estimate the projection in the image of the center of the landmark. Figure 1.5 illustrates the case where the landmark occupies relatively few pixels, which is the most challenging case and a common one in image measurement applications, because space is limited in both the image and in the scene in which the landmark lies. The measured intensity levels, one quantized value per pixel, are the data available for landmark location estimation.

Automated assembly and inspection, vision-guided robotic manipulation, photogrammetry, image metrology and camera calibration all benefit from precise landmark location, and so it is useful to know how accurately artificial landmarks can be located digital images. It is well known that subpixel accuracy can be achieved; but to answer the question of whether a landmark can be located with an accuracy of ± 0.1, 0.01 or 0.001 pixels requires detailed consideration. This work aims to answer the question of landmark location precision by establishing the theoretical minimum uncertainty that any location estimator can achieve, by developing tools for evaluating the uncertainty of practical estimators, and by presenting an experiment for directly measuring landmark location precision. The fruits of these labors can be used to determine the accuracy of a specified system, or to engineer a system to meet an accuracy goal. As O'Gorman *et al.* point out,

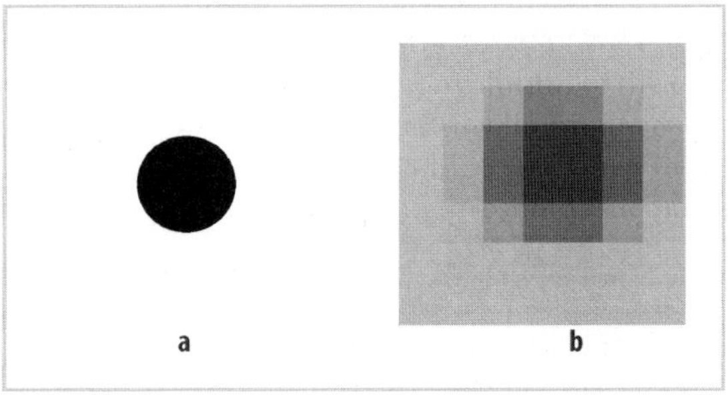

Figure 1.5a,b. Circular landmark; **a** ideal image of the landmark, **b** digital image of the landmark. Each picture element of the digital image corresponds to a measured intensity level (or three measured intensities for color images).

"… the benefits of methods that achieve subpixel precision can be seen from two perspectives: either as a means to obtain a higher precision, or as a means to obtain the same precision at less computational cost." (O'Gorman *et al.* 1990).

Since the present work is quite narrowly focused, it is worth mentioning what the present work is not. It is not a study of the entire range of questions that arise when measuring from images. In particular, the question of how a landmark is detected is neglected. For automated image processing this is an important and challenging question, but for the present purposes it is assumed that the presence of the landmark in the image is known, along with its approximate location. Furthermore, the sources of uncertainty considered are restricted to those that arise with the imaging and estimation processes, which is to say that any uncertainty in the shape of the landmark itself is neglected. This is effectively a restriction to well-made artificial landmarks. Indeed, in the numerical and experimental studies of the present report only planar-circular landmarks are considered, although the methods extend to other shapes. Here, the only concern is with the location of the landmark in the image, and so, although the numerical tool incorporates a lens distortion model, displacement of the landmark in the image due to lens distortion does not affect the conclusions. Finally, only digital images are considered; that is, images that are quantized both in space (pixels) and intensity. This is more a challenge than a simplification, since the main theoretical challenge of this work is to deal with quantization processes when determining the uncertainty. Narrowly focused as it is, the present work does have these chief aims:

1. To introduce a measure for landmark location uncertainty that overcomes the nonstationary statistics of the problem.

2. To develop an extended model for the physics of image formation, capturing aspects of the physical process that contribute to the determination of the uncertainty.

3. To establish the Cramér–Rao lower bound (CRLB) for the uncertainty of landmark location, given a description of the landmark and imaging configuration. The CRLB establishes a minimum level of uncertainty for any unbiased estimator, independent of the algorithm used.

4. To build a set of tools for analyzing the uncertainty of practical landmark and imaging configurations and location estimation algorithms, taking into account the spatial and intensity quantization in the image and the nonstationary nature of the statistics in particular.

5. To demonstrate new, model-based landmark location estimators that perform near the Cramér–Rao theoretical limit.

6. To validate these tools by developing and executing an experiment that can directly measure the uncertainty in landmark location estimates. The principal challenge to experimental validation is to provide a means to separately and accurately establish the true location of a landmark in the image.

These six goals form the basis for the remainder of the work.

1.1 Prior Art

Landmark location uncertainty has been studied for many years by numerous investigators of photogrammetry and machine vision. The methods used by scientists to approach landmark location uncertainty problems can be divided into two major categories: statistical and geometrical.

The first approaches used statistical approximation for the quantization and sampling errors and then applied well-established statistical techniques (Förstner 1982; Dvornychenko 1983). Gonsalves (1976), Ho (1983), Ryan *et al.* (1982), and Sriraman *et al.* (1989) modeled the entire digital image as a stochastic process so that statistical analysis could be used to determine the bounds for the error variances.

As a variation on the statistical approaches, several authors explored the use of autocorrelation peak searching with known landmark templates. Tian and Hunhns (1986) studied correlation interpolation and non-noisy intensity interpolation on synthetic images; that is, images created artificially following a deterministic mathematical model. They concluded that the major factors affecting landmark location uncertainty are the choice of interpolation function (based on the model used), the spatial sampling density (resolution), and the amplitude quantization level (gray level). The drawback of this methodology is that it requires a descriptive model of the image to be analyzed, not just the landmark. Nishihara and Crossley (1988) developed and implemented a similar technique that was applied to the alignment of photolithographic images in integrated circuit manufacturing. The authors analyzed the effects of noisy images on the location uncertainty. In this case, the application features a controlled environment where the model of the image (size, shape, and orientation) is known in advance.

Similarly, Cox *et al.* (1990) addressed the problem of subpixel landmark location for surface-mounted printed circuit board (PCB) assembly operation by developing an algorithm that used the model of an image as a reference for a 2-D matching search to predict the error in the measurement. A variation on the above methods was introduced by Tichem and Cohen (1994), where the centroid and model of a circularly symmetric landmark was used as a seed for the matching algorithm. This study recognized the difficulty of landmark location when the landmark is not coplanar with the imager, so that the landmark image is altered by perspective distortion.

Parametric modeling of the landmark was used initially by Duda and Hart (1972), Shapiro (1978), Yam and Davis (1981), and Ballard (1981), with emphasis placed on landmark detection via the use of Hough transforms and little attention paid to the landmark location uncertainty problem. Several other authors, such as Thurgood and Mikhail (1982) and Hachicha *et al.* (1989), used parametric modeling of particular landmark models to infer landmark location uncertainty from statistical properties of the parameter estimates. This technique was applied successfully by Tabatabai and Mitchell (1984) and Nalwa and Binford (1986) in the precision edge location application.

In contrast to statistical methodologies, analytic techniques used in landmark location have focused on geometric approaches. The first generation of these

Table 1.1. Mapping of relevant publications vs. key elements for landmark location uncertainty

		Landmark Geometry	Intensity Quantization	Imager Noise	Smoothing Effects
Statistical Modeling	Widrow (1965)		×		
	Duda and Hart (1972)				
	Gonsalves (1976)		×		
	Shapiro (1976)				
	Ballard (1981)				
	Ryan *et al.* (1982)				
	Förstner (1982)				
	Thurgood and Mikhail (1982)	×			
	Dvornychenko (1983)				
	Ho (1983)				
	Doros (1984)	×			
	Nielsen *et al.* (1984)		×		
	Tabatabai and Mitchell (1984)	×			
	Nalwa and Binford (1986)	×			×
	Bruckstein (1987)		×		
	Nishihara and Crossley (1988)				
	Hachicha *et al.* (1988)				
	Sriraman *et al.* (1989)				
	Tian and Huhns (1990)				
	Cox *et al.* (1990)				
	Tichem and Cohen (1994)	×			

Table 1.1. (continued)

		Landmark Geometry	Intensity Quantization	Imager Noise	Smoothing Effects
Geometric Approaches	Klassman (1975)	×			
	Kulpa (1979)	×			
	Hyde and Davis (1983)	×	×		
	Nakamura and Aizawa (1984)	×			
	Dorst and Smeulders (1986)	×			
	Berenstein *et al.* (1987)	×			
	Havelock (1989)	×	×		
	Amir (1990)	×			
	O'Gorman *et al.* (1990)	×			
	Havelock (1991)	×	×	×	
	Bose and Amir (1990)	×			
	Kiryati and Bruckstein (1991)	×	×		
	Chiorboli and Vecchi (1993)	×	×		
	Efrat and Gotsman (1994)	×			
	O'Gorman (1996)	×			
	Bruckstein *et al.* (1998)	×			
	Bruckstein and Holt (1999)	×			×

Table 1.2. Summary of research evolution for landmark location uncertainty

	Binary Digital Image				Gray-level Digital Image			
Method	Lines	Circles	Other Shape (e.g. arc)	General	Lines	Circles	Other Shape (e.g. arc)	General
Statistical: Statistical Modeling Correlation Methods Parametric Modeling	Duda and Hart (1972)							Widrow (1965)
				Gonsalves (1976)				
				Shapiro (1978)				
	Yam and Davis (1981)			Ballard (1981)				
				Förstner (1982)				
				Tabatabai and Mitchell (1984)				
			Thurgood and Mikhail (1982)	Ryan *et al.* (1982)				
				Dvornychenko (1983)				
				Ho (1983)				
								Nielsen *et al.* (1984)
					Nalwa and Binford (1986)			
								Bruckstein (1987)
				Nishihara and Crossley (1988)			Hachicha *et al.* (1989)	
				Sriraman *et al.* (1989)				
				Cox *et al.* (1990)				
		Tichem and Cohen (1994)						

Table 1.2. (continued)

Method	Binary Digital Image				Gray-level Digital Image			
	Lines	Circles	Other Shape (e.g. arc)	General	Lines	Circles	Other Shape (e.g. arc)	General
Geometric	Klaasman (1975)							
		Kulpa *et al.* (1979)						
	Vossepoel and Smeulders (1982)							
								Hyde and Davis (1983)
		Nakamura and Aizawa (1984)	Doros (1984)					
	Dorst and Smeulders (1986)							Tian and Huhns (1986)
	Berenstein *et al.* (1987)	Overington (1987)						
								Havelock (1989)
		Amir (1990)						
	Bose and Amir (1990)	Bose and Amir (1990)						Havelock (1991)
		O'Gorman *et al.* (1990)						
								Kiriaty and Bruckstein (1991)
					Chiorboli and Vecchi (1993)	Chiorboli and Vecchi (1993)	Chiorboli and Vecchi (1993)	
		Efrat and Gotsman (1994)						
				O'Gorman (1996)				
				Bruckstein *et al.* (1998)				
				Bruckstein and Holt (1999)				

techniques was based on binary digital images, disregarding gray levels. Kulpa (1979) provided an in-depth geometrical analysis of the properties of digital circles, but Hill (1980) was the first to provide a rigorous geometrical analysis of landmark location uncertainty for a variety of geometrical shapes based on synthetic binary images. Subsequently, several investigators reported a variety of algorithms for improving error performance for particular landmark shapes. Vossepoel and Smeulders (1982), Dorst and Smeulders (1986), and Berenstein *et al.* (1987) focused on calculating uncertainty in the location of straight spatially quantized lines. Nakamura and Aizawa (1984) investigated the expansion of this work to circular landmarks, while Amir (1990) and Efrat and Gotsman (1994) presented new methodologies for locating the center of a circular landmark and Doros (1986) worked on the geometrical properties of the image of digital arcs. Bose and Amir (1990) investigated how landmark location uncertainty is affected by the shapes and sizes of the binary images of various landmarks, concluding that circular shapes provide superior performance. Similarly, O'Gorman (1990), O'Gorman *et al.* (1996), and Bruckstein *et al.* (1998) investigated novel shapes (bullseye and others), which improved the uncertainty performance for binary images.

The second generation of analytic techniques focused on the exploitation of the gray-level information, taking advantage of more bits in the intensity quantization. The use of grayscale intensity quantization improves landmark location uncertainty (Rohr 2001). One of the first algorithms to make use of grayscale information was developed by Hyde and Davis (1983), which gave results similar to the ones obtained for binary images. Klassman (1975) proved that for any finite spatial sampling density (density of pixels per area), the position uncertainty of a line was greater than zero, even when using unquantized images; that is, images whose amplitude is a continuous function. Kiryati and Bruckstein (1991) showed how location uncertainty performance on binary image could be improved by the use of gray-level digitizers. Chiorboli and Vecchi (1993) showed empirically that the uncertainty performance of the results from Bose and Amir could be improved by an order of magnitude by using gray-level information.

Of particular importance is the locale framework proposed by Havelock (1989, 1991) and expanded by Zhou *et al.* (1998). This methodology is based on the concept of regions of indistinguishable object position, called locales, caused by spatial and intensity quantization. In the absence of noise, an observed digital (quantized) image of a landmark corresponds to a region in which the true landmark may lie. In general, this is a many-to-one mapping; that is, many different landmark locations generate the same digital image. Havelock's work marked a great leap forward in analytic determination of landmark location uncertainty.

Table 1.1 shows how selected research publications address aspects of image formation that are important for landmark location uncertainty. A summary of the research evolution for landmark location uncertainty is shown in Table 1.2. As the table reflects, research has been directed toward understanding and reducing landmark location uncertainty for over three decades.

1.2 Modeling Image Formation

In this section, the approach of this work to modeling image formation is described. The goal is to capture the physical processes of image formation (such as properties of the camera and imager) that are important when determining the ultimate limits to landmark location uncertainty.

While many previous studies have focused on binary images, several investigators have recognized the need to develop methods for subpixel landmark location that make more efficient use of the information present in the image. Havelock (1989), a pioneer in the investigation of landmark location uncertainty, points out that "it is not trivial to answer these questions [of landmark location uncertainty with grayscale information] in a complete and meaningful way. This is a new area of investigation, with only few results available at present." Similarly, the work of Bruckstein *et al.* (1998) recommends further research into the exploitation of the gray-level information to improve precision in landmark location.

Grayscale information is only one of the characteristics that influence landmark location uncertainty. An extensive model of image formation comprises contributions from several scientific fields: physics (optics), electronics (imager), and computer science (image processing and estimation algorithms). The research framework presented here introduces a methodology to determine the minimal uncertainty bound in landmark location based on the Cramér–Rao lower bound. The CRLB methodology considers an extensive model of image formation, including six degree-of-freedom (DOF) landmark to camera geometry, diffraction, defocus, lens distortion, grayscale, pixel geometry, and pixel sensitive area. A list is shown in Figure 1.6 of the aspects of the image formation process that are considered in this work. The Cramér–Rao lower bound methodology provides a theoretical statistical minimum limit on the landmark location uncertainty. Knowledge of this bound provides the means to evaluate the actual performance of both existing and future landmark location estimators. With this model, a general framework is established to measure the uncertainty performance of existing landmark location estimators, *e.g.*, binary, geometrical, autocorrelation, statistical modeling, and others.

An important challenge to addressing uncertainty in landmark location is that the statistics of landmark location are not stationary. That is to say, the theoretical CRLB as well as the covariance and bias of practical estimators all depend on *where* the landmark falls on the imager. This characteristic is associated with the locales and raises the question of how to express uncertainty, since the covariance matrix itself varies significantly with small shifts of the true landmark location. A method is needed which averages over the range landmark locations and yet conveys information of practical engineering significance. At the same time, the method chosen to express uncertainty should be applicable to all three of the measures sought: the Cramér–Rao lower bound, analysis of practical estimators, and experimental measurement.

The confidence interval has been chosen to express landmark location uncertainty. The confidence interval is expressed as the radius of a circle in the digital image that will hold 95% (or in some cases 99%) of estimated landmark locations.

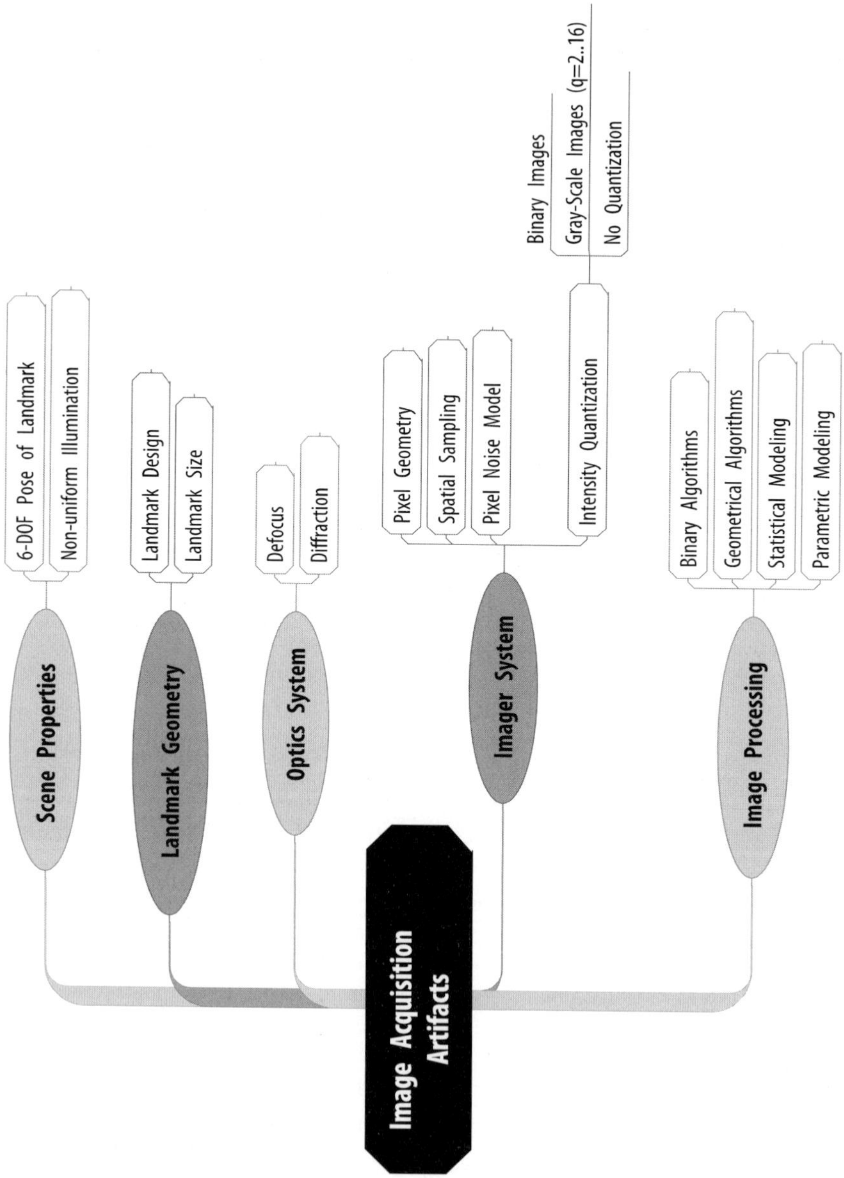

Figure 1.6. Contributors to the uncertainty bound in landmark location

In Chapter 3, methods for evaluating the confidence interval for the CRLB and analytic cases are described in detail. For the experimental data, the confidence intervals are obtained simply by determining the smallest circle that will enclose 95% of the data. While coordinates of millimeters or microns on the imager are preferred over pixel coordinates for image metrology, we have chosen to express the confidence interval in millipixels because landmark location estimators operate on pixel data. It is recognized that a circle measured in pixels translates, in general, to an ellipse in image coordinates.

Using the confidence intervals, results are summarized by means of performance maps. These maps are a graphical representation of the location uncertainty computed using the CRLB, or the analysis of a practical algorithm, or as measured by experiment. Each performance map corresponds to a particular set of model parameters, called a configuration. Figure 1.7 shows two examples of hypothetical landmark location estimation performance maps corresponding to two different configurations.

The landmark location estimation performance maps permit identification of the gap between existing algorithms and the minimal uncertainty bound. Furthermore, they pave the way for the creation of novel estimators that approach the bound by identifying the main image formation factors that contribute to the uncertainty bound.

The complexity of an extensive physical model of image formation, involving the characteristics shown in Figure 1.6, has been a barrier to the study of landmark location uncertainty. As we will see in the next chapter, to correctly determine the interaction of small displacements of the landmark and the corresponding small changes in the measured data, it is necessary to consider the finite sensitive

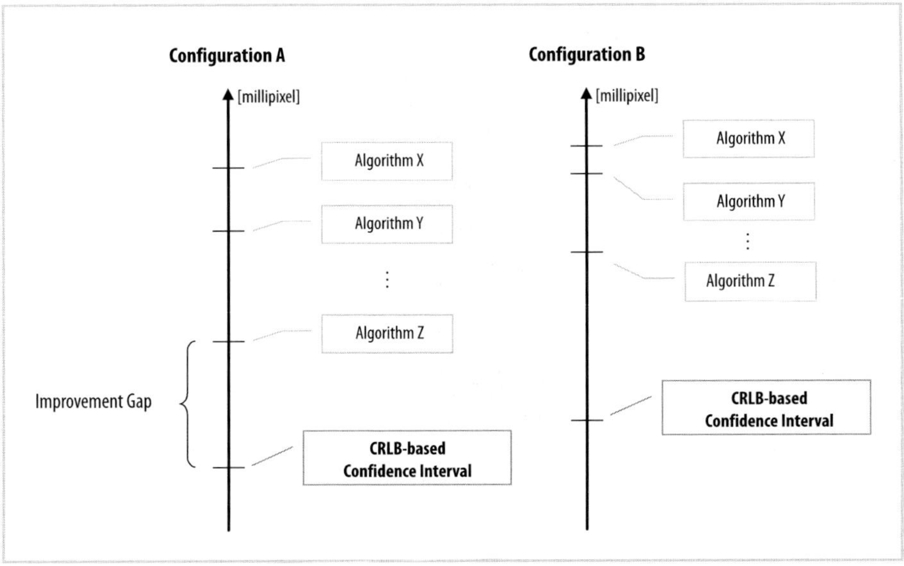

Figure 1.7. Hypothetical landmark location estimation performance maps

area of each pixel and also the smoothing of the image introduced by diffraction. Since the sensitivity of the image data to the landmark location is essential to this study, neither characteristic can be neglected. However, modeling finite sensitive area requires integration of the incident light intensity (called the illuminance function) over the two-dimensional sensitive area of each pixel, and this integral cannot be performed in closed form. Indeed, because of the 2-D convolution required to model diffraction-induced smoothing of the illuminance function, the illuminance function itself cannot be expressed in closed form for any case other than a straight edge.

The strategy used here to address these challenges employs a mix of analysis, where feasible, and numerical work where required. The inclusion of diffraction in the analysis of landmark location uncertainty is an important innovation, because the smoothing of the illuminance function by diffraction *enables* rigorous numerical analysis. In particular, the partial derivatives of the probability distributions of the measured data with respect to the location of the landmark, which are needed to compute the CRLB and also the performance of practical estimators, are well defined, well behaved, and can be computed only because diffraction assures us that the illuminance function has no discontinuities. This motivates the incorporation of diffraction as part of the model of image formation.

This work also introduces two novel circular landmark location estimation algorithms, which perform with confidence intervals at the tens of millipixels level. This level of uncertainty is several times lower than that of a more conventional centroid algorithm. These algorithms are demonstrated for circular landmarks, but the model-based methodology can be extended to accommodate other landmark designs.

The first new landmark location estimator is named the "ellipsoidal contour" algorithm. It exploits the gray-level information present in the image by identifying points on a grayscale contour around the landmark, and then estimating the parameters of an ellipse fit to the contour. The second estimator, named "Butterworth tepuy", also exploits gray-level information by creating a parametric model of the intensity surface of the landmark, while considering the smoothing caused by diffraction.

To experimentally measure the uncertainty of a landmark location algorithm, it is necessary to have an independent measurement of the true location in the image of the landmark. This challenge is met by designing a second artwork that surrounds and is much larger than the primary landmark. By averaging over hundreds of pixels along the boundaries of this artwork, a reference location is determined for evaluating location estimators applied to the much smaller primary landmark. The design incorporates an internal measure of location accuracy, to verify that the larger secondary landmark has been located more accurately than the Cramér–Rao lower bound for the location uncertainty of the primary landmark.

In summary, the present work establishes—for the first time—the minimum uncertainty in landmark location attainable, considering an extensive physical model of image formation. With this bound, landmark location performance maps are created to show the gap between the uncertainty achieved by practical landmark location methodologies and the uncertainty bound. Moreover, novel landmark location estimation algorithms are introduced that achieve uncertainty levels that closely approach the Cramér–Rao lower bound. Finally, experimental results are presented to validate the theoretical findings.

1.3 Mathematical Symbols and Nomenclature

The discussion of landmark location accuracy will involve details of geometry expressed in several coordinate frames. A notation that captures the type of representation, the point referred to and the coordinate frame of reference will provide a compact way to discuss landmark geometry unambiguously. The notation used by J.J. Craig (1986) and others is described here and used throughout this work. Notation related to signals and estimation is also described. In addition to this section, a list of symbols is provided in Appendix A.

1.3.1 Coordinate Systems

Point b expressed in coordinate system A is given by ${}^{A}P_b$. The coordinate systems used will be:

"w": World coordinates ${}^{w}P_b \in \Re^3 = [{}^{w}X_b \quad {}^{w}Y_b \quad {}^{w}Z_b]^T$, *i.e.*, user-defined inertial coordinate system

"c": Camera coordinates ${}^{c}P_b \in \Re^3$, *i.e.*, coordinate system centered at the principal point of the camera/lens; the z-axis ${}^{c}Z$ lies along the principal ray of the lens

"L": Landmark coordinates, *i.e.*, coordinate system embedded in a landmark. Two-dimensional landmarks are considered here, so ${}^{L}P_b \in \Re^2$

"i": Image coordinates ${}^{i}P_b \in \Re^2$, *i.e.*, a metric coordinate system centered at the intersection of the principal ray and the imager

"p": Pixel coordinates ${}^{p}P_b \in \Re^2$, *i.e.*, a coordinate system co-planar with the image coordinate system, centered in the upper-left corner of the imager

Generally, points expressed in image or pixel coordinates are points in the image. For example, a point b in a landmark has coordinates ${}^{L}P_b$ given by the mathematical description of the landmark, ${}^{w}P_b$ given by ${}^{L}P_b$ and the transformation between landmark and world coordinates, and point ${}^{i}P_b$ is the image of point b expressed in millimeters on the imager, while ${}^{p}P_b$ is the pixel coordinates of the same image point. Some additional examples are:

${}^{w}P_b$ point "b" in world coordinates

$$ {}^{w}P_b = \begin{bmatrix} {}^{w}X_b \\ {}^{w}Y_b \\ {}^{w}Z_b \end{bmatrix} = {}^{w}\begin{bmatrix} X \\ Y \\ Z \end{bmatrix}_b \tag{1.1} $$

${}^{c}P_b$ point "b" in camera coordinates

$$ {}^{c}P_b = \begin{bmatrix} {}^{c}X_b \\ {}^{c}Y_b \\ {}^{c}Z_b \end{bmatrix} = {}^{c}\begin{bmatrix} X \\ Y \\ Z \end{bmatrix}_b \tag{1.2} $$

${}^{L}P_a$ point "a" in landmark coordinates

$$ {}^{L}P_a = \begin{bmatrix} {}^{L}X_a \\ {}^{L}Y_a \end{bmatrix} = {}^{L}\begin{bmatrix} X \\ Y \end{bmatrix}_a \quad (1.3)$$

1.3.2 Origin of the Coordinate System

The origin is a special point in a coordinate frame. The origin of coordinate frame A expressed in coordinate frame D is denoted as ${}^{D}P_{\underset{A}{o}}$. The upper-left subscript may be omitted when a general reference to the origin of the coordinates system is made. Several examples are:

Center of a landmark in camera coordinates:

$$ {}^{c}P_{\underset{L}{o}} = \begin{bmatrix} {}^{c}X_{\underset{L}{o}} \\ {}^{c}Y_{\underset{L}{o}} \\ {}^{c}Z_{\underset{L}{o}} \end{bmatrix} = {}^{c}\begin{bmatrix} X \\ Y \\ Z \end{bmatrix}_{\underset{L}{o}} \quad (1.4)$$

Origin of the image coordinates:

$$ P_{\underset{i}{o}} \quad (1.5)$$

The image location of the center of the landmark, expressed in pixel coordinates:

$$ {}^{p}P_{\underset{L}{o}} \quad (1.6)$$

${}^{a}\mathcal{P}_b$ Symbol ${}^{a}\mathcal{P}_b$ designates a six-dimensional pose of a coordinate frame b expressed in coordinate frame a. An example used in the development is ${}^{c}\mathcal{P}_{\underset{L}{o}}$, which is the pose of the landmark expressed in camera coordinates:

$$ {}^{c}\mathcal{P}_{\underset{L}{o}} = \begin{bmatrix} {}^{c}x & {}^{c}y & {}^{c}z & \omega & \kappa & \varphi \end{bmatrix}^T , \quad (1.7)$$

where $\begin{bmatrix} {}^{c}x & {}^{c}y & {}^{c}z \end{bmatrix}^T$ is the 3-DOF position of the landmark in the camera frame, and where ω = pitch, κ = roll and φ = yaw angles of the orientation of the landmark in the camera frame. For this work, the rotation order for the rotation from target to camera coordinates is $R = R_\kappa R_\varphi R_\omega$, where each R is a 3×3 rotation matrix.

1.3.3 Image Formation

The following subsection describes the nomenclature and mathematical symbols that will be used at each stage of the image formation model. The image formation model is described in detail in Chapter 2.

Landmark luminosity function:

Function that provides the definition of the landmark geometry on the landmark plane,

$$f_L({}^L x, {}^L y)\,, \tag{1.8}$$

expressed in lumens or per unit.

Optical flux function:

Function representing the spatial distribution of illumination of the imager considering geometric optics:

$$H({}^i x, {}^i y) = \Phi\left(G_f\left({}^c\mathcal{P}_{\underset{L}{o}}, f_L({}^L x, {}^L y)\right)\right)\,, \tag{1.9}$$

expressed in lux or per unit.

Illuminance function:

Function representing the optical flux on the imager after the addition of the diffraction phenomena:

$$E({}^i x, {}^i y) = \Xi\left(H({}^i x, {}^i y)\right)\,, \tag{1.10}$$

expressed in lux or per unit.

Discrete analog intensity function:

Function describing the illuminance function after spatial quantization:

$$J(i,j) = Q_s\left(E({}^i x, {}^i y)\right) \tag{1.11}$$

expressed in charge or per unit.

Digital image:

Function describing the discrete analog intensity function after amplitude quantization:

$$I = Q_i(J(i,j))\,, \tag{1.12}$$

expressed in units of counts on a range determined by the A/D converter, or per unit with 1 = full scale.

1.3.4 Estimation Basics

Estimation is the statistical discipline that involves estimating the values of model parameters based on measured data. Much of the present work employs the tools of a multivariable statistical framework with a focus on estimation theory. The following mathematical notation is used throughout.

The true value of a given parameter vector θ is marked with a star:

$$\theta^* \,. \tag{1.13}$$

The estimated value of a given parameter θ is marked with a hat:

$$\widehat{\theta} \,. \tag{1.14}$$

The misadjustment in the estimate of a give parameter θ is marked with a tilde: $\widetilde{\theta}$; that is,

$$\widehat{\theta} = \theta^* + \widetilde{\theta} \,. \tag{1.15}$$

The measurement of a data vector Y is written as: $\overline{Y}$. For each measurement $\overline{Y}$, there is an underlying but generally unknown true value, Y^*, and measurement error or noise, $\widetilde{Y}$, with a relationship:

$$\overline{Y} = Y^* + \widetilde{Y} \,. \tag{1.16}$$

Examples:

- Estimated center of landmark in image coordinates:

$$\widehat{{}^{i}P_{\underset{L}{o}}} \,. \tag{1.17}$$

- Estimated location of a point b in pixel coordinates:

$$\widehat{{}^{p}P_b} = {}^{p}P_b^* + \widetilde{{}^{p}P_b} \,, \tag{1.18}$$

where $\widetilde{{}^{p}P_b}$ represents estimator error.

- Measured digital image in the presence of noise:

$$\overline{I} = Q_I(\overline{J}) = Q_I(J^* + \widetilde{J}) \,, \tag{1.19}$$

where $\widetilde{J}$ is noise in the discrete analog intensity function and $Q_I(\cdot)$ is the intensity quantization function.

- Noise in the measurement of the digital image:

$$\widetilde{I} = I^* - \overline{I} \,. \tag{1.20}$$

1.3.5 Estimators

The landmark location estimators are algorithms that take image data and produce a landmark location estimate, written ${}^{p}\widehat{P}_{0} = T\left(\bar{I}\right)$. Four landmark location estimators are considered; these are denoted by $T_A()$, $T_B()$, $T_D()$, and $T_E()$, where:

$T_A()$ is the binary centroid estimator;
$T_B()$ is the grayscale centroid estimator;
$T_D()$ is the ellipsoidal contour estimator, and;
$T_E()$ is the Butterworth tepuy estimator.

Some nomenclature examples for estimators follow:

- Binary centroid estimator on the discrete analog intensity function J:

$$ {}^{p}\widehat{P}_{\underset{L}{o}} = T_A(J) \,. \tag{1.21} $$

- Butterworth tepuy estimator applied to the digital image function I:

$$ {}^{p}\widehat{P}_{\underset{L}{o}} = T_E(I) \,. \tag{1.22} $$

- Estimate of the center of the landmark in pixel coordinates using the ellipsoidal estimator:

$$ {}^{p}\widehat{P}_{\underset{L}{o}} = T_D(I) \,. \tag{1.23} $$

1.4 Content Organization

The subsequent chapters are organized as follows. Chapter 2 addresses modeling issues, with the extensive physical model of digital image formation developed in Section 2.1. Locales and the use of confidence intervals are discussed in Section 2.2. The necessary analytic tools are developed in Chapter 3, with tools for forming the confidence interval based on the Cramér–Rao lower bound derived in Section 3.1 and tools for analyzing the performance of practical estimators derived in Section 3.2. The details of two new, model-based landmark location estimators are presented in Chapter 4. Calculation of both the CRLB and practical estimator performance requires an innovation in 2-D numerical integration, which is the topic of Chapter 5. The computational tool that generates the landmark location uncertainty performance maps is presented in Chapter 6. In Chapter 7 an experimental method is presented which can directly and accurately measure the uncertainty in locating a primary landmark, based on the estimated location of a much larger and more accurate secondary landmark. This is followed in Chapter 8 by the presentation of experimentally measured landmark location uncertainty, which is used to validate the analytic tools of Chapter 3, as well as the results from numerical investigations employing the analytic tools to study the impact on location accuracy of a range of configuration variables. Finally, the conclusions of this work are presented in Chapter 9.

A downloadable MATLAB® package to assist the reader with applying theoretically-derived results to practical engineering configurations is available from http://www.springer.com/978-1-84628-912-5.

2 Physics of Digital Image Formation

We begin our discussion of modeling by addressing why it is essential to consider two phenomena that play a role in image formation, influence the precision of landmark location, and have often been neglected in the past. These are the finite sensitive area of each pixel and the smoothing of the illuminance function introduced by diffraction. The detection of landmark location is an estimation problem, and the accuracy of any estimator is governed by the sensitivity of the data to changes in the quantity to be estimated. For landmark location, this is $\left(\partial J/\partial\,{}^{p}P_{o}^{*}\right)$, where $J \in \Re^{N_p \times 1}$ is the discrete analog intensity values of N_p pixels used to estimate the landmark location, and ${}^{p}P_{o}^{*} \in \Re^{2}$ is the true location of the landmark. The intensity values are organized as an $N_p \times 1$ column vector so that the derivative $\left(\partial J/\partial\,{}^{p}P_{o}^{*}\right)$ is a 2-D array. For this discussion, the analog intensity values, J, are used rather than the digital image values, I, to avoid the complication of intensity quantization. Consideration of intensity quantization is deferred to Section 3.2.1.

A hypothetical ideal image of a landmark is illustrated in Figure 2.1, which shows a 3×3 pixel grid with a two-pixel-diameter circular landmark. The white point in each pixel indicates a simplified model of image formation in which each pixel is sampled at its center to determine the measurements of the digital image. For example, referring to the indices shown in parentheses in Figure 2.1, the digital image is given as:

$$J = [0.0\ \ 0.0\ \ 0.0\ \ 0.0\ \ 1.0\ \ 1.0\ \ 0.0\ \ 1.0\ \ 1.0]^{T}\ , \tag{2.1}$$

where a value of $J(i) = 1.0$ corresponds to a point inside the landmark image, and $J(i) = 0.0$ corresponds to a point in the background. With this model, which combines the point-wise sensitive area of the pixel with a discontinuous boundary for the landmark image, 0.0 and 1.0 are the only possible values of pixel intensity. This simplified model has been used in 36 of the 38 articles listed in Table 1.1.

An example showing the sensitivity of the data to changes in landmark location is illustrated in Figures 2.2 and 2.4. These figures schematically show the circular landmark and 3×3 grid of pixels of Figure 2.1. Like Figure 2.1, Figure 2.2 is drawn with pixels modeled as having a sensitive area that is 0% of the pixel area, and no smoothing at the edge of the landmark. In the left portion of Figure 2.2, the landmark is shown at location ${}^{p}P_{1}^{*} = [2.45\ \ 2.40]^{T}$, and in the right portion it is at location ${}^{p}P_{1}^{*} = [2.45\ \ 2.35]^{T}$. Ordering the data in a column vector according to

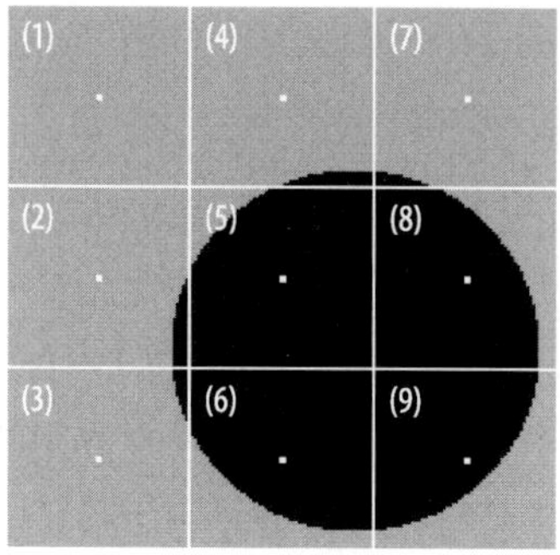

Figure 2.1. Illustration of the optical flux function corresponding to a circular landmark, shown on a digital imager. The *squares* indicate pixels, which are indexed by the numerals in parentheses. An example optical flux function, $H({}^ix, {}^iy)$, is shown, see Equation 1.9 above. Because the landmark has a sharp edge, the optical flux is discontinuous at the boundary of the landmark. The *white points* in the centers of the pixels indicate 0% sensitive area.

indices shown in parentheses gives

$$J_1 = \begin{bmatrix} 0.0 \\ 0.0 \\ 0.0 \\ 0.0 \\ 1.0 \\ 1.0 \\ 0.0 \\ 1.0 \\ 1.0 \end{bmatrix} \qquad J_2 = \begin{bmatrix} 0.0 \\ 0.0 \\ 0.0 \\ 0.0 \\ 1.0 \\ 1.0 \\ 0.0 \\ 1.0 \\ 1.0 \end{bmatrix} \qquad \Delta J / \Delta {}^PY_o^* = \begin{bmatrix} 0.0 \\ 0.0 \\ 0.0 \\ 0.0 \\ 0.0 \\ 0.0 \\ 0.0 \\ 0.0 \\ 0.0 \end{bmatrix}, \qquad (2.2)$$

where $\Delta J / \Delta {}^PY_o^*$ is the change in image data arising with the change in landmark location. In this example, even though the landmark has moved between the left and right portions of Figure 2.2, the data are not modified. When finite sensitive

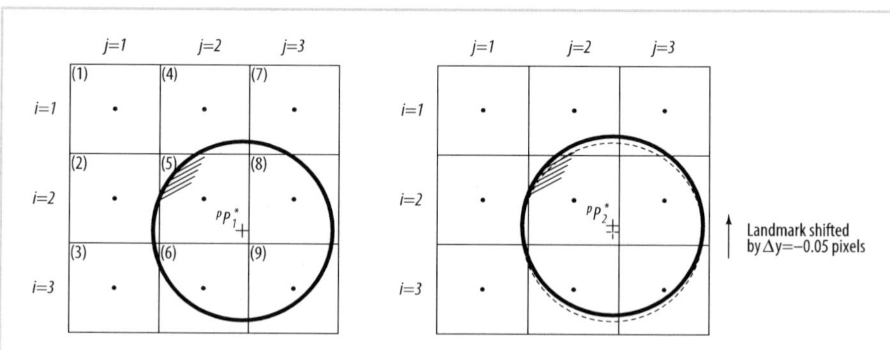

Figure 2.2. A circular landmark on a 3 × 3 pixel grid. The pixel data is organized into a column vector according to the indices shown in parentheses. The interior of the landmark is shown with *cross-hatching*. The center of the landmark is shown with a *cross* and marked ${}^PP_1^*$. The landmark covers the sensitive areas of pixels 5, 6, 8 and 9.

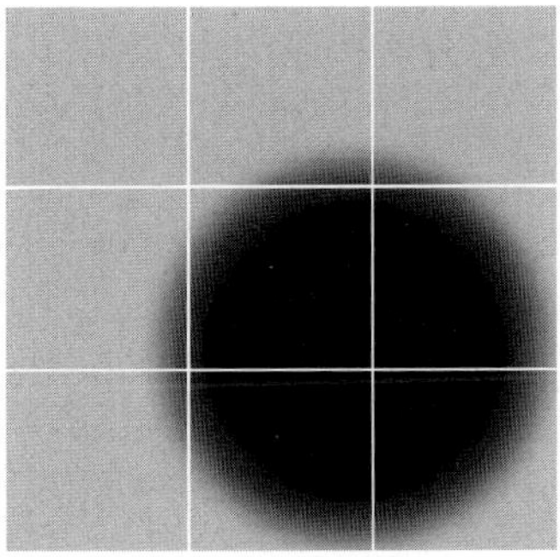

Figure 2.3. Illustration of a landmark image with smoothing considered. An example illuminance function, $E({}^ix, {}^iy)$, is shown (see Equation 1.10 above). The smoothing effect of diffraction is approximated by convolving the intensity profile of Figure 2.1 with a Gaussian kernel. The illuminance function is continuous throughout the image, and in particular, at the boundary of the landmark. Pixel sensitive area is not shown.

area and smoothing are not considered, the model predicts an analog intensity function that is a piece-wise constant, discontinuous function of the landmark location; and the derivative of the data with respect to the landmark location is either zero or infinity, either of which pose difficulties when examining estimator uncertainty.

Figure 2.3 illustrates a simulated image of the same landmark seen in Figure 2.1, except that diffraction is incorporated into the model. To generate the illustration of Figure 2.3, diffraction is approximated by convolution with a 2-D Gaussian kernel, as described in Section 2.1. Because of the smoothing effect of diffraction, the intensity of the image of the landmark is not discontinuous.

In Figure 2.4, a small displacement of the landmark is once again considered. Figure 2.4 is like Figure 2.2, except that the pixel sensitive area is modeled as 12% of the total pixel area, and the shaded ring illustrates a region of smooth transition between the interior and exterior intensity of the landmark. In the transition region shown by the shaded ring, the illuminance function takes intermediate

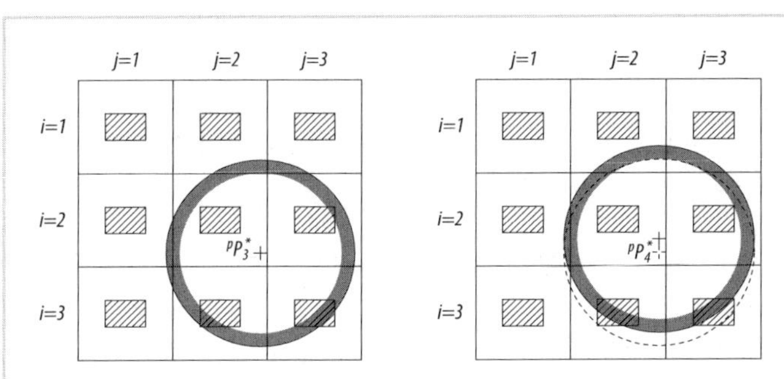

Figure 2.4. A circular landmark on a 3×3 pixel grid with finite sensitive area. This figure is analogous to the previous figure, but with finite sensitive areas and transition region shown. Pixel (2,2) will measure an illuminance corresponding fully to the landmark, while three other pixels will record partial illumination by the landmark.

values between $E({}^{i}x, {}^{i}y) = 1.0$ and $E({}^{i}x, {}^{i}y) = 0.0$. On the right-hand side, as in Figure 2.2, the landmark is shifted by 0.05 pixels. Example data for Figure 2.4, along with the first difference of the data with respect to the ${}^{p}Y$ position of the landmark, are given as

$$J_3 = \begin{bmatrix} 0.0 \\ 0.0 \\ 0.0 \\ 0.0 \\ 1.00 \\ 0.90 \\ 0.00 \\ 0.95 \\ 0.70 \end{bmatrix} \qquad J_2 = \begin{bmatrix} 0.0 \\ 0.0 \\ 0.0 \\ 0.0 \\ 1.00 \\ 0.60 \\ 0.00 \\ 1.0 \\ 0.30 \end{bmatrix} \qquad \Delta J/\Delta {}^{p}Y_o^* = \begin{bmatrix} 0.0 \\ 0.0 \\ 0.0 \\ 0.0 \\ 0.00 \\ 0.60 \\ 0.00 \\ -1.00 \\ 0.80 \end{bmatrix}, \tag{2.3}$$

The nonzero values of the first difference show that there is a signal with which to detect the change of location of the landmark. The first difference above relates to the derivatives that are developed theoretically in Section 3.1.1 and computed with the tools described in Chapter 6.

There are several issues that are illustrated by the examples of Figure 2.1 through Figure 2.4 that should be noted:

- With the simplified model of Figure 2.2, the data do not contain any indication of a small shift in the landmark location. These figures show an oversimplification, because actual imagers have finite sensitive area—in most cases greater than 12% fill factor—and actual illuminance distributions have finite transition regions.
- Because diffraction gives a smooth transition from background to landmark illuminance levels, the derivative of the data with respect to the landmark location is mathematically well posed. Smoothing makes the calculation of the Cramér–Rao lower bound possible.
- The pixel boundaries are artificial. It is the boundaries of the sensitive areas that have physical significance and determine the connection between landmark configuration and measured data. For convenience, the pixel boundaries are set so that the centroid of the pixels coincide with the centroid of the sensitive areas.
- The derivative of the image data with respect to displacements of the landmark depends on the location of the landmark. This gives rise to the nonstationary statistics of landmark location estimation. These nonstationary statistics are treated in Sections 3.1.2 for the Cramér–Rao lower bound, and 3.2.2 for the analysis of practical estimators.

With these considerations, the formation of the digital image of a landmark can be modeled as a set of transformations that start with the geometrical definition of the landmark and lead to a spatially quantized and intensity-quantized image array $I(i, j)$, or digital image. Image formation can be modeled in five stages:

1. First, the 6-DOF pose of the landmark in camera coordinates is passed through a landmark description transformation $G_f\left({}^c\mathcal{P}_{\underset{L}{\circ}}, f_L\left({}^Lx, {}^Ly\right)\right)$, creating the landmark luminosity function $f_L({}^cx, {}^cy, {}^cz)$.
2. The landmark luminosity function is then transformed by the geometric optics transfer function Φ, creating the *optical flux* function $H(x,y)$. A pinhole camera model is used (Hecht and Zajac 1979). A sample optical flux function is shown in Figure 2.1.
3. The *optical flux* function is then low-pass-filtered by the transfer function Ξ, creating the *illuminance function* $E(x,y)$. The low-pass filtering models the diffraction of the light passing through the camera aperture, as predicted by

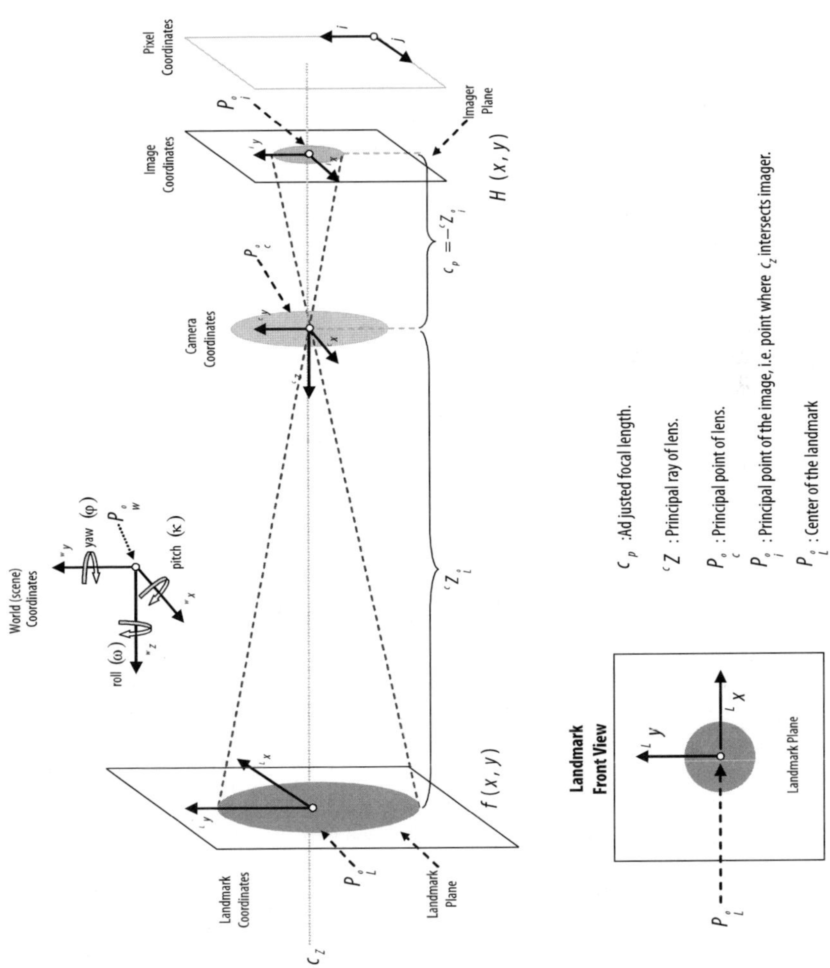

Figure 2.5. Geometric Optics and Coordinate System Definition

Fourier optics. An example illuminance function, corresponding to the optical flux function of Figure 2.1, is illustrated in Figure 2.3.

4. The *Illuminance function* is processed by the image transducer, or imager, which performs a spatial sampling transformation Q_s, yielding the *discrete analog intensity* function $J(i,j)$.
5. Finally, the amplitude of the discrete analog intensity function is quantized by the transformation Q_i, creating the *digital image*, $I(i,j)$.

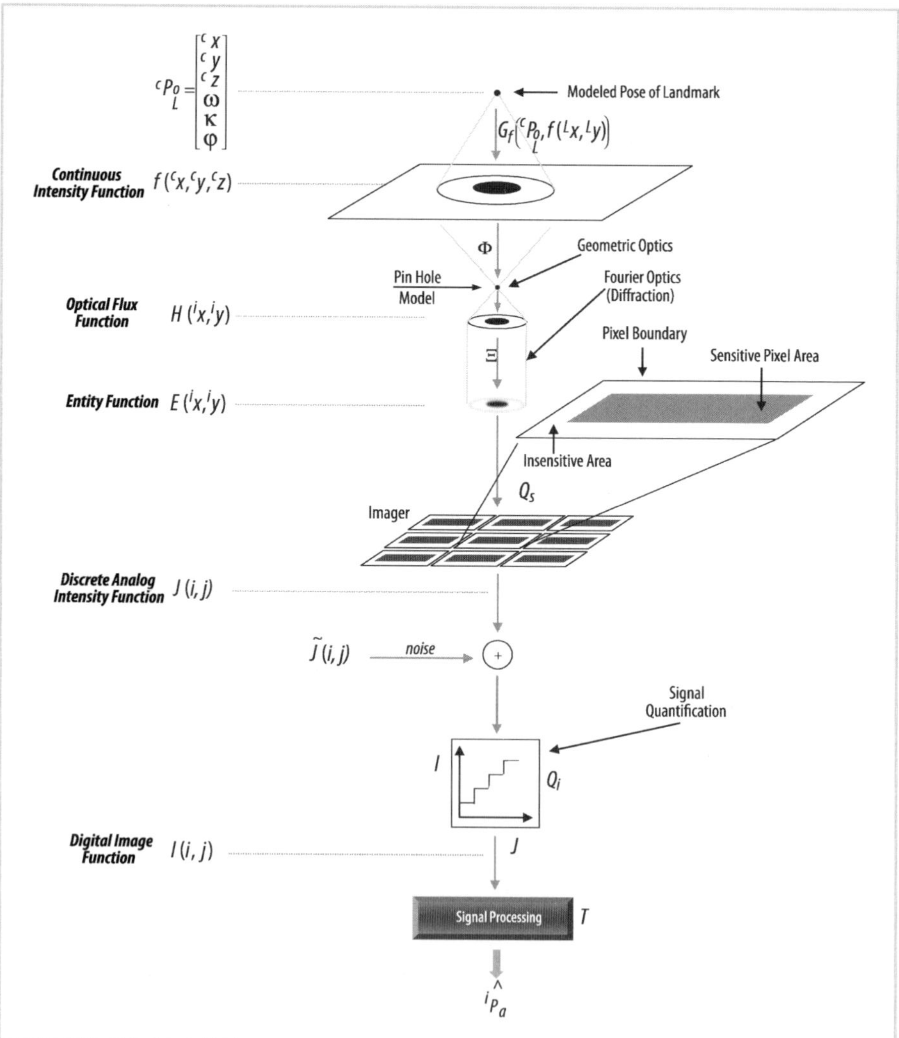

Figure 2.6. Generation of a digital image

The coordinate frames of the geometric optical model are illustrated in Figure 2.5. Figure 2.6 shows the model chain of transformations. The mathematical nomenclature used in Equation 2.4 is presented in Section 1.3 above, and is listed in Appendix A.

The model chain of transformations is shown in functional form in Equation 2.4. The last transformation presented in Equation 2.4 corresponds to the estimation of the location of the landmark.

$$
{}^{c}\mathcal{P}_{\underset{L}{o}} = \begin{bmatrix} {}^{c}x \\ {}^{c}y \\ {}^{c}z \\ \omega \\ \kappa \\ \varphi \end{bmatrix} \overset{G_f}{\rightarrow} f({}^{L}x, {}^{L}y) \overset{\Phi}{\rightarrow} H({}^{i}x, {}^{i}y) \overset{\Xi}{\rightarrow} E({}^{i}x, {}^{i}y) \overset{Q_s}{\rightarrow} J({}^{P}i, {}^{P}j) \rightarrow
$$

$$
\rightarrow \underset{\widetilde{J}({}^{P}i, {}^{P}j)}{\underset{\uparrow}{\oplus}} \overset{Q_i}{\rightarrow} I({}^{P}i, {}^{P}j) \overset{T}{\rightarrow} {}^{i}\widehat{P}_{\underset{L}{o}} \,. \tag{2.4}
$$

Notice that image noise $\widetilde{J}(x, y)$ is added at the stage just before the intensity quantization; physically this noise model is consistent with the properties of practical

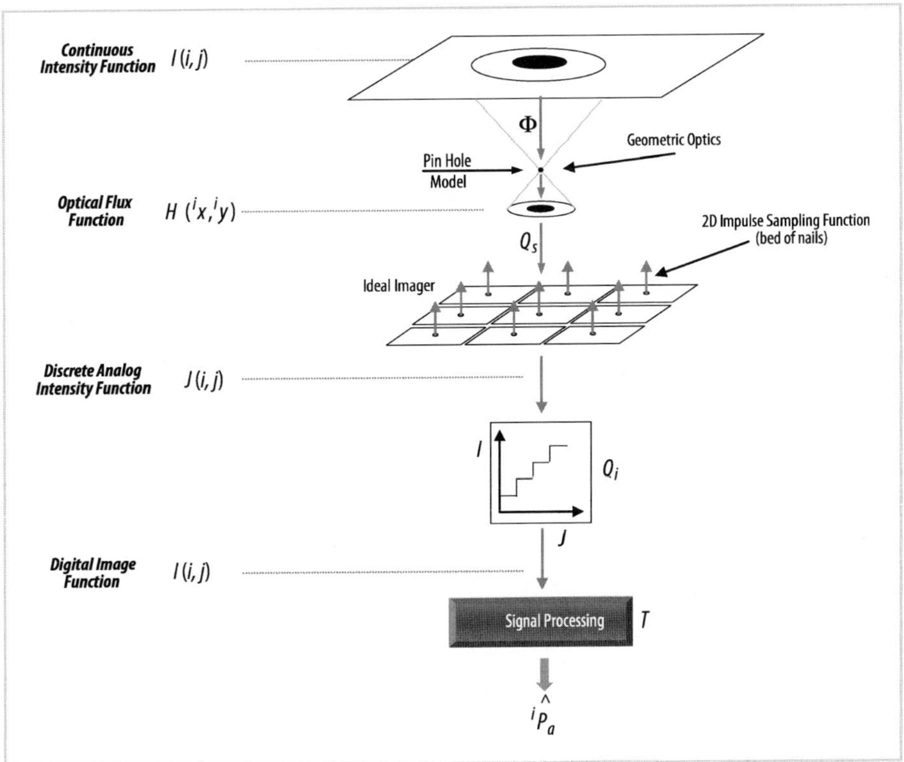

Figure 2.7. Simplified model of digital image formation used in past studies

imagers. Furthermore, in this model, as in practical image acquisition systems, it is assumed that the optical flux function is of finite support; that is, the region Ω in which it is nonzero has finite extent; this is expressed mathematically in Equation 2.5.

$$H(x,y) = 0 \quad \{(x,y) \notin \Omega\} . \tag{2.5}$$

In the previous studies of landmark location uncertainty listed in Tables 1.1 and 1.2, it was common to assume a discontinuous transition in the illuminance function at the edge of the image of the landmark, and to treat the spatial sampling process as impulsive; that is, to model the photosensitive area of the pixel as a point in the sampling plane. For comparison with Equation 2.4 and Figure 2.6, Equation 2.6 and Figure 2.7 show the simplified model of digital image formation that has been used in many previous studies.

$$F({}^{L}x, {}^{L}y) \overset{\Phi}{\rightarrow} H({}^{i}x, {}^{i}y) \overset{Q_s}{\rightarrow} J(i,j) \overset{Q_i}{\rightarrow} I(i,j) . \tag{2.6}$$

2.1 Image Formation and Landmark Location Uncertainty

Landmark location uncertainty is affected by a variety of artifacts in the image acquisition process. As indicated in Figure 1.6, these artifacts can be classified into five groups: scene properties, landmark geometry, optics system, imager system, and the selection of the image-processing algorithm. The following paragraphs provide an overview of how these artifacts participate in the overall uncertainty in landmark location.

2.1.1 Scene Properties

The two main scene properties that affect the landmark location process are the six degree-of-freedom (6-DOF) pose of landmark and nonuniform illumination. When the landmark is tilted with respect to the optical axis of the camera, it will give rise to perspective distortion in the image of the landmark. Nonuniform illumination can introduce bias into the location estimator.

The effects of perspective distortion can be visualized by means of a simple example. Figure 2.8 illustrates the influence of landmark tilt. In this figure, portions of the landmark that are closer to the camera appear larger than portions farther away from it, causing the overall shape of the landmark to be distorted. Consequently, the centroid of the image of the landmark no longer corresponds to the image of the centroid of the landmark. Full perspective distortion is considered here through the use of analytic tools developed in this work. Landmark location errors arising with tilt are investigated in Section 8.6.

Nonuniform illumination causes intensity variations in the image that may cause bias in the image-processing algorithm. Landmark location errors that arise with tilt are investigated in Section 8.8.

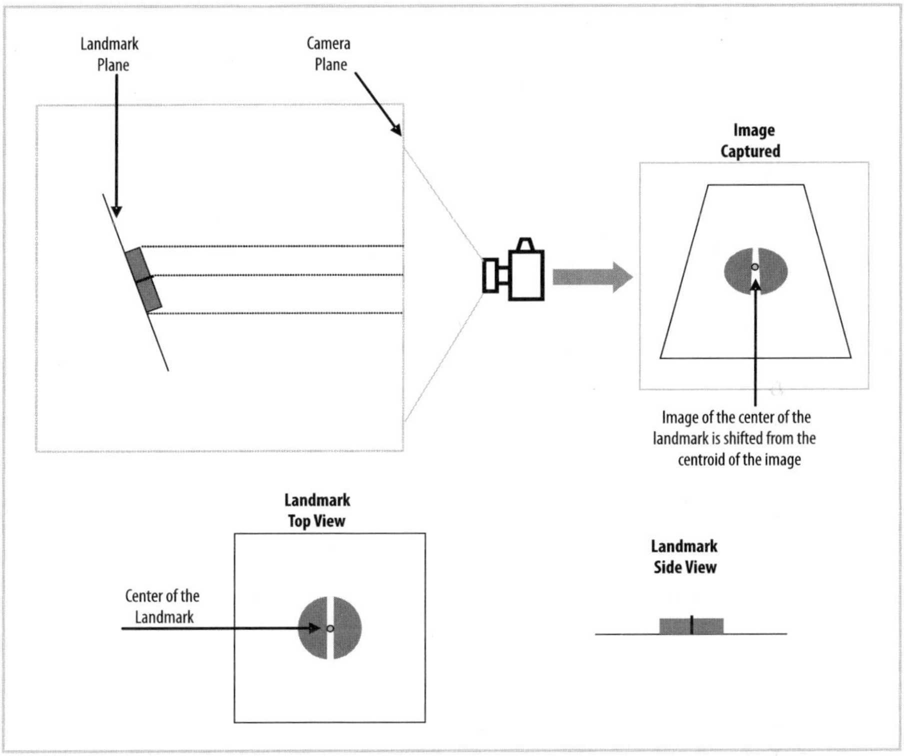

Figure 2.8. Example of the effects of perspective distortion in landmark location

2.1.2 Landmark Geometry

The landmark shape directly impacts on the landmark location uncertainty (Bose and Amir 1990; Amir 1990; O'Gorman *et al.* 1990). Normally, the impact on location uncertainty depends on the type of image processing algorithm (Bruckstein *et al.* 1998). Typical landmark shapes used are shown in Figure 2.9. Usually, landmarks with circular symmetry are preferred due to their rotational invariance and analytically simple formulation (Bose and Amir 1990).

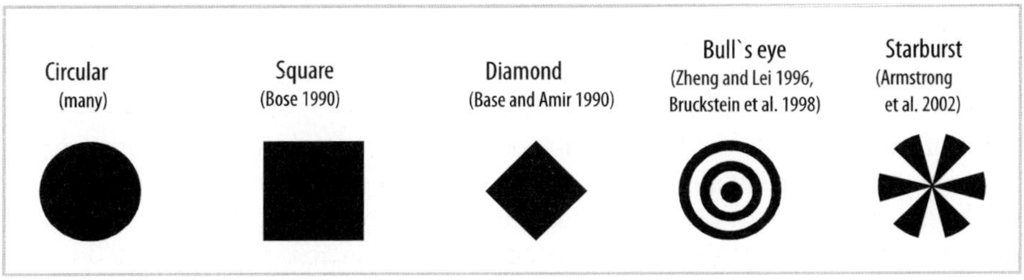

Figure 2.9. Typical landmark shapes used in the literature

Similarly, the size of the landmark with respect to the scene captured effects the location uncertainty. The landmark size determines the number of pixels carrying information that the image-processing algorithm can exploit.

2.1.3 Optics System

The optical system of the camera forms the flux function of the scene on the imager sensor. Optical phenomena such as diffraction and defocus influence the uncertainty of landmark location.

Diffraction

Diffraction is the phenomenon that occurs when light deviates from straight-line propagation in a way that cannot be interpreted as reflection or refraction. Diffraction arises when light waves are partially blocked by obstacles (Hecht and Zajac 1979). Optical devices, such as cameras, collect a limited portion of the incident light wavefront from the scene; this inevitably introduces diffraction (Sears 1979), which is perceived as smoothing of the image. The degree of smoothing is a function of the geometry of the camera aperture and its size relative to λ, the wavelength of light being captured. In the conceptual limit when $\lambda \to 0$, the diffraction effect no longer exists and all of the ideal geometrical optics laws apply.

A model of diffraction is incorporated into this study for two reasons: the smoothing of the illuminance function by diffraction influences the performance of landmark location estimators, and it also makes the required numerical integration of the illuminance function possible. Without smoothing the illuminance function, the analytic results of this study would not be possible.

The diffraction pattern caused by a circular aperture is called the Airy pattern (Figure 2.10) and is described by:

$$I(\theta) = I(0) \left[\frac{2J_1\left(k\frac{D}{2}\sin\theta\right)}{k\frac{D}{2}\sin\theta} \right]^2 , \tag{2.7}$$

where $I(0)$ is the maximum illuminance at the center of the pattern, D is the diameter of the circular aperture, $k = \frac{2\pi}{\lambda}$ is the propagation number, J_1 in the equation above is the Bessel function of the first kind and order 1, and θ is the angle formed between the principal ray and a vector from the principal point toward a point (x, y) in the image plane; that is, $\tan(\theta) = \frac{\sqrt{{}^ix^2 + {}^iy^2}}{c_p}$. Justification of this equation comes from the Fraunhofer approximation to the Huygens–Fresnel principle and the use of circular apertures (Goodman 1968).

The first dark ring of the Airy pattern has a flux density of about 1.75% that of the central ring. About 84% of the light incident on the image is distributed in the main lobe of the Airy pattern. The central disk of this pattern is called the Airy

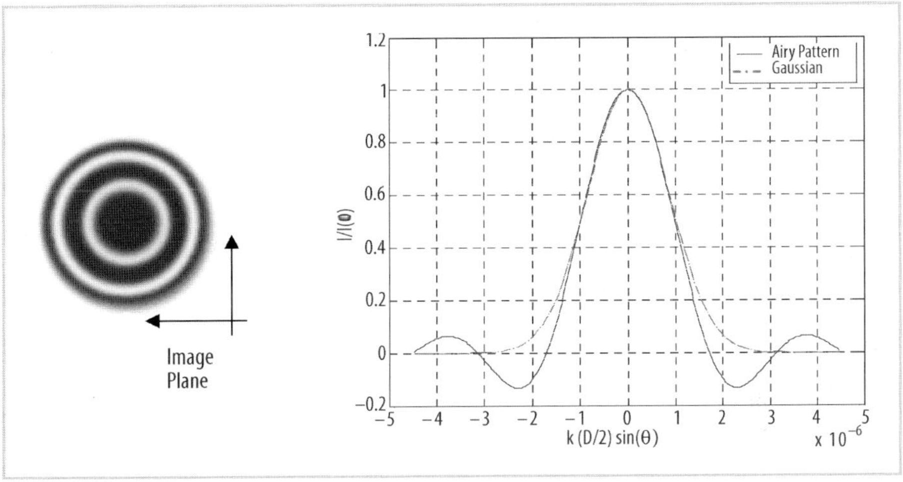

Figure 2.10. The Airy pattern produced by a point source diffracted by a circular aperture. The Gaussian curve closely approximates the central lobe.

disk. The radius r_a of the Airy disk is given by (Hecht and Zajac 1979):

$$r_a = 1.22 \frac{c_p}{D} \lambda \,, \tag{2.8}$$

where c_p is the principal distance, or the distance from the imager plane to the lens principal point, $c_p \approx f$ where f is the lens focal length, D is the diameter of the lens aperture, and λ is the wavelength of the light being diffracted.

The presence of the Airy pattern diminishes the resolving power of optical systems. To address this issue, camera manufacturers use a process called apodization. Apodization is a process that diminishes the side lobes of the diffraction pattern by altering the shape of the aperture or modifying its transmission properties (Hecht and Zajac 1979). Even without the use of apodization by design, nonlinearities and imperfections in the optical system (*e.g.*, defocus) suppress the side lobes of the Airy pattern.

Here, the pattern resulting from apodization is approximated with a Gaussian function with standard deviation σ_b. As seen in Figure 2.10, the side lobes are suppressed, even though the main lobe may be broadened. Following this approximation, diffraction will be modeled as the convolution of the optical flux function $H(x, y)$ and a 2-D Gaussian kernel, to produce the illuminance function $E(x, y)$ (see Equation 2.4 and Figure 2.6); that is,

$$E(x, y) = \frac{1}{2\pi\sigma_b^2} \mathrm{e}^{-\frac{x^2+y^2}{2\sigma_b^2}} * H(x, y) \,, \tag{2.9}$$

where $*$ represents the 2-D convolution operator (Gonzales and Woods 1992). Relating Equation 2.9 to the example at the start of this chapter, Figure 2.1 illustrates the optical flux function, $H(x, y)$, and Figure 2.3 illustrates the illuminance function, $E(x, y)$. For the approximation of Figure 2.10, the 2-D Gaussian kernel is related to the Airy pattern by

$$\sigma_b = \frac{r_a}{2} . \tag{2.10}$$

The radius of the Airy disk can also be related to camera parameters. By definition, the focal number, $F/\#$, is given as the relation between the lens focal length and the diameter of the camera aperture; that is,

$$F/\# = \frac{f}{D} . \tag{2.11}$$

The wavelength range of visible light is:

$$\lambda = 780 \ldots 390 \text{ nm} . \tag{2.12}$$

For $\lambda = 780$ nm, we get a useful value for the radius of the smoothing operator of:

$$r_a = 1.22 F/\#\lambda \approx F/\# \text{ (in microns)} . \tag{2.13}$$

This implies that the radius of the Airy disk can be approximated by the $F/\#$ in microns. Using Equation 2.10 and Equation 2.13 we get:

$$\sigma_b = \frac{1}{2} F/\# \text{ (in microns)} . \tag{2.14}$$

For practical systems, the radius of the smoothed pattern given as the image of a small point should be measured and related to the radius of a 2-D Gaussian distribution. Furthermore, some image acquisition systems specify the parameter σ_b.

Defocus

Departures from the diffraction-limited optical model are called aberrations (IEEE 2000). The most common optical aberration is caused by defocus. The lens law determines the focus condition:

$$\frac{1}{f} = \frac{1}{c_\mathrm{p}} + \frac{1}{{}^c z_{\underset{L}{o}}} , \tag{2.15}$$

where f is the focal length of the lens, c_p is the distance from the principal point of the lens to the imager, sometimes called the adjusted focal length, and ${}^c z_{\underset{L}{o}}$ is the distance from the lens to the observation plane, as indicated in Figure 2.11.

Defocus creates a blurring effect on the image, which can be modeled by a suitable change in the convolution operator of Equation 2.9. The convolution operator to model defocus is the image of the suitably scaled aperture (commonly a uniform disk). Using Fourier optics, it can be shown that defocus lowers the contrast of each spatial frequency component of image intensity (Goodman 1968). Defocus increases the smoothing provided by diffraction. When the dimension of the smoothing operator σ_b is small relative to the pixel size, additional smoothing can

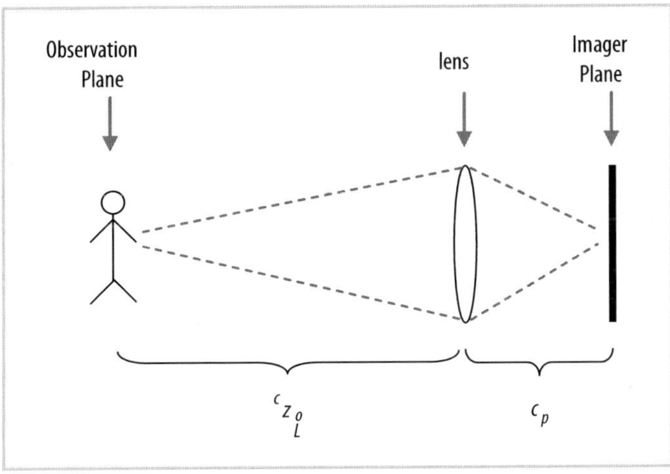

Figure 2.11. Geometry of the lens law, Equation 2.15

enhance location accuracy. Thus, there may be cases in which it is advantageous to deliberately defocus a camera for precision landmark location.

The tools developed in Chapters 3 and 6 support detailed analysis of landmark location uncertainty with specific levels of smoothing. Since the convolution of Equation 2.9 cannot be performed in closed form, it is implemented numerically using the specialized 2-D integration technique presented in Chapter 5 to generate the illuminance function for a particular configuration. Because of the information lost in the regions between pixel sensitive areas, as seen in Figure 2.4, smoothing of the digital image by signal processing is not equivalent to smoothing of the illuminance function by defocus.

Lens Distortion

Lens distortion is a type of lens aberration that alters the location of the image of the landmark (Atkinson 1996). There are two types of lens distortions: radial and decentering.

Radial distortion is caused by the curvature of the lens and consists of a radial displacement of the image of the landmark closer to or farther from the principal point. Radial distortion is normally expressed in the form known as balanced ∂r_b, as:

$$\partial r_b = K_0 r + K_1 r^3 + K_2 r^5 + K_3 r^7 + \cdots \text{ [microns]} \tag{2.16}$$

where the constants K_i represent the coefficients of radial distortion and r is the radial distance from the principal point:

$$r^2 = \left({}^{i}x - {}^{i}x_o\right)^2 + \left({}^{i}y - {}^{i}y_o\right)^2 \qquad \left({}^{i}x_o, {}^{i}y_o\right) = \text{Principal Point.} \tag{2.17}$$

In many cases the terms K_0 and K_1 are sufficient. For wide-angle lenses, higher order terms may also be significant.

Decentering distortion is caused by any displacement or rotation of a lens element from perfect alignment. This type of aberration is caused by manufacturing defects, and it causes geometric displacement in images. The effects of decentering distortion are approximated in a truncated polynomial form as:

$$\Delta x_S = \left(1 - \frac{c_p}{{}^c z_{\underset{L}{o}}}\right) \left[P_1 \left(r^2 + 2\left({}^i x - {}^i x_o\right)^2\right) + 2P_2 \left({}^i x - {}^i x_o\right)\left({}^i y - {}^i y_o\right)\right]$$

$$\Delta y_S = \left(1 - \frac{c_p}{{}^c z_{\underset{L}{o}}}\right) \left[P_2 \left(r^2 + 2\left({}^i y - {}^i y_o\right)^2\right) + 2P_1 \left({}^i x - {}^i x_o\right)\left({}^i y - {}^i y_o\right)\right], \quad (2.18)$$

where Δx_S, Δy_S represents the decentering distortion at an image point ${}^i x, {}^i y$; r is the radial distance as described in Equation 2.17, $\left({}^i x_o, {}^i y_o\right)$ are the coordinates of the principal point, and c_p is the adjusted focal length for a lens focused at a distance ${}^c z_{\underset{L}{o}}$.

The distortion model of Equation 2.17 and Equation 2.19 is implemented in the computation tools described in Chapter 6. In addition, the lens used in the experimental section shown in Chapter 7 was calibrated by determining the lens distortion parameters using a commercial tool.

2.1.4 Imager System

Several artifacts of the imager system participate in the determination of landmark location uncertainty. An overview of these artifacts follows.

The first artifact considered is spatial sampling. The 2-D sampling theorem is discussed. Whereas the sampling theorem provides that the band-limited sampled signal can be completely reconstructed, the spatial frequency of the illuminance function is rarely, if ever, limited to the spatial frequency bandwidth described by the sampling theorem. The significant consequence of this is a skew error introduced by spatial sampling, which appears as a bias in the estimate landmark location. An example with impulsive spatial sampling (infinitesimal pixel sensitive area) and a 1.8-pixel-diameter landmark illustrates the bias.

Next finite sensitive area is considered, along with a model for noise in the discrete analog intensity function. Finally, intensity quantization is considered. With these pieces in place, locales—regions in which the landmark can lie which give the same digital image—are considered in detail.

Spatial Sampling and Locales

The *Whittaker–Shannon* sampling theorem states that to recover a band-limited, continuous signal of bandwidth W from its samples without distortion, it is necessary to select a sampling interval Δx so that $1/\Delta x \geq 2W$ (Ziemmer *et al.* 1983), or

$$\Delta x \leq \frac{1}{2W}, \quad (2.19)$$

where Δx is the width of a pixel on the imager (in millimeters), and W (in cycles/mm) is the spatial bandwidth of the illuminance function. The sampling theorem can be expanded for two-dimensional signals. In this case, the 2-D sampling theorem states that a continuous band-limited function $E\left({}^{i}x, {}^{i}y\right)$ can be recovered completely from samples whose separation is bounded by

$$\Delta x \leq \frac{1}{2W_x}\,, \tag{2.20}$$

$$\Delta y \leq \frac{1}{2W_y}\,, \tag{2.21}$$

where W_x and W_y represent the bandwidths of the signal in the x and y coordinates respectively. The estimate coordinates $(\widehat{x}, \widehat{y})$ of the unsampled illuminance function can be calculated estimating the centroid of the landmark; that is,

$$\widehat{x} = \frac{\iint\limits_{\Omega} x\,E(x,y)\,dx\,dy}{\iint\limits_{\Omega} E(x,y)\,dx\,dy} = \frac{\int\limits_{y_l}^{y_u}\int\limits_{x_l}^{x_u} xE(x,y)\,dx\,dy}{\int\limits_{y_l}^{y_u}\int\limits_{x_l}^{x_u} E(x,y)\,dx\,dy}$$

and

$$\widehat{y} = \frac{\iint\limits_{\Omega} y\,E(x,y)\,dx\,dy}{\iint\limits_{\Omega} E(x,y)\,dx\,dy} = \frac{\int\limits_{y_l}^{y_u}\int\limits_{x_l}^{x_u} yE(x,y)\,dx\,dy}{\int\limits_{y_l}^{y_u}\int\limits_{x_l}^{x_u} E(x,y)\,dx\,dy}\,. \tag{2.22}$$

From the *Whittaker–Shannon* sampling theorem, the same centroid estimate could be obtained from the samples of the band-limited signal, since the original unsampled function could be reconstructed from its samples from $-\infty$ to ∞. However, an image function, just like all physical signals, is necessarily of finite extent, which in turn implies that it cannot be band-limited and its samples will only be able to approximate the original function.

An estimate of the centroid based on the samples of the illuminance function can be defined in the following manner:

$$\widehat{x}_d = \frac{\sum\limits_{i=0}^{\frac{x_u-x_l}{\Delta x}} \sum\limits_{j=0}^{\frac{y_u-y_l}{\Delta y}} \left(x_l + i\Delta x\right) E(x_l + i\Delta x,\ y_l + j\Delta y)}{\sum\limits_{i=0}^{\frac{x_u-x_l}{\Delta x}} \sum\limits_{j=0}^{\frac{y_u-y_l}{\Delta y}} E(x_l + i\Delta x,\ y_l + j\Delta y)}$$

and

$$\widehat{y}_d = \frac{\sum_{i=0}^{\frac{x_u - x_l}{\Delta x}} \sum_{j=0}^{\frac{y_u - y_l}{\Delta y}} (y_l + j\Delta y)\, E(x_l + i\Delta x,\ y_l + j\Delta y)}{\sum_{i=0}^{\frac{x_u - x_l}{\Delta x}} \sum_{j=0}^{\frac{y_u - y_l}{\Delta y}} E(x_l + i\Delta x,\ y_l + j\Delta y)}. \tag{2.23}$$

The estimate $(\widehat{x}_d, \widehat{y}_d)$ represents the discrete centroid of the illuminance function $E(x, y)$. The difference of this estimate from the actual position is referred to as the *skew error*; this error is caused by the spatial sampling process.

It can be shown that in the limit as the sampling interval goes to zero, the discrete centroid estimate is an unbiased estimator of the unsampled illuminance function centroid; that is, $\Delta x \to 0$ and $\Delta y \to 0$ implies that $(\widehat{x}_d, \widehat{y}_d) \to (\widehat{x}, \widehat{y})$. The proof follows from the integral definition applied to Equation 2.23:

$$\operatorname*{Lim}_{\Delta x \to 0} \operatorname*{Lim}_{\Delta y \to 0} \left(\widehat{x}_d\right) = \operatorname*{Lim}_{\Delta x \to 0} \operatorname*{Lim}_{\Delta y \to 0} \frac{\Delta x \Delta y \sum_{i=0}^{\frac{x_u - x_l}{\Delta x}} \sum_{j=0}^{\frac{y_u - y_l}{\Delta y}} (x_l + i\Delta x)\, E(x_l + i\Delta x,\ y_l + j\Delta y)}{\Delta x \Delta y \sum_{i=0}^{\frac{x_u - x_l}{\Delta x}} \sum_{j=0}^{\frac{y_u - y_l}{\Delta y}} E(x_l + i\Delta x,\ y_l + j\Delta y)}$$

$$= \frac{\operatorname*{Lim}_{\Delta x \to 0} \operatorname*{Lim}_{\Delta y \to 0} \sum_{i=0}^{\frac{x_u - x_l}{\Delta x}} \sum_{j=0}^{\frac{y_u - y_l}{\Delta y}} x_i E(x_i, y_i)\, \Delta x \Delta y}{\operatorname*{Lim}_{\Delta x \to 0} \operatorname*{Lim}_{\Delta y \to 0} \sum_{i=0}^{\frac{x_u - x_l}{\Delta x}} \sum_{j=0}^{\frac{y_u - y_l}{\Delta y}} E(x_i, y_i)\, \Delta x \Delta y} = \frac{\int_{y_l}^{y_u} \int_{x_l}^{x_u} x E(x, y)\, dx\, dy}{\int_{y_l}^{y_u} \int_{x_l}^{x_u} E(x, y)\, dx\, dy} = \widehat{x}\,;$$

similarly,

$$\operatorname*{Lim}_{\Delta x \to 0} \operatorname*{Lim}_{\Delta y \to 0} \left(\widehat{y}_d\right) = \frac{\int_{y_l}^{y_u} \int_{x_l}^{x_u} y E(x, y)\, dx\, dy}{\int_{y_l}^{y_u} \int_{x_l}^{x_u} E(x, y)\, dx\, dy} = \widehat{y}\,. \tag{2.24}$$

For image processing, the requirements of the sampling theorem are generally not met, and thus skew error is introduced into the centroid location estimator. To illustrate the skew error in landmark location caused by 2-D spatial sampling, consider centroid estimation of the location of the circular landmark shown in the subfigures of Figure 2.12. In this example, the landmark has a diameter of 1.8 pixels and the illuminance of the imager is without smoothing, that is $\sigma_b = 0$. As a further simplification, impulsive sampling is considered; that is, with a landmark intensity of A and a background intensity of 0, the discrete analog intensity of each pixel is either A or 0, depending on whether the sensitive area of the pixel (shown as a dot in Figure 2.12) lies within or outside the landmark. The small landmark size is used to work with a 3×3 patch of pixels. It is clear from the figure that pixels outside this patch play no role when estimating the location of the landmark.

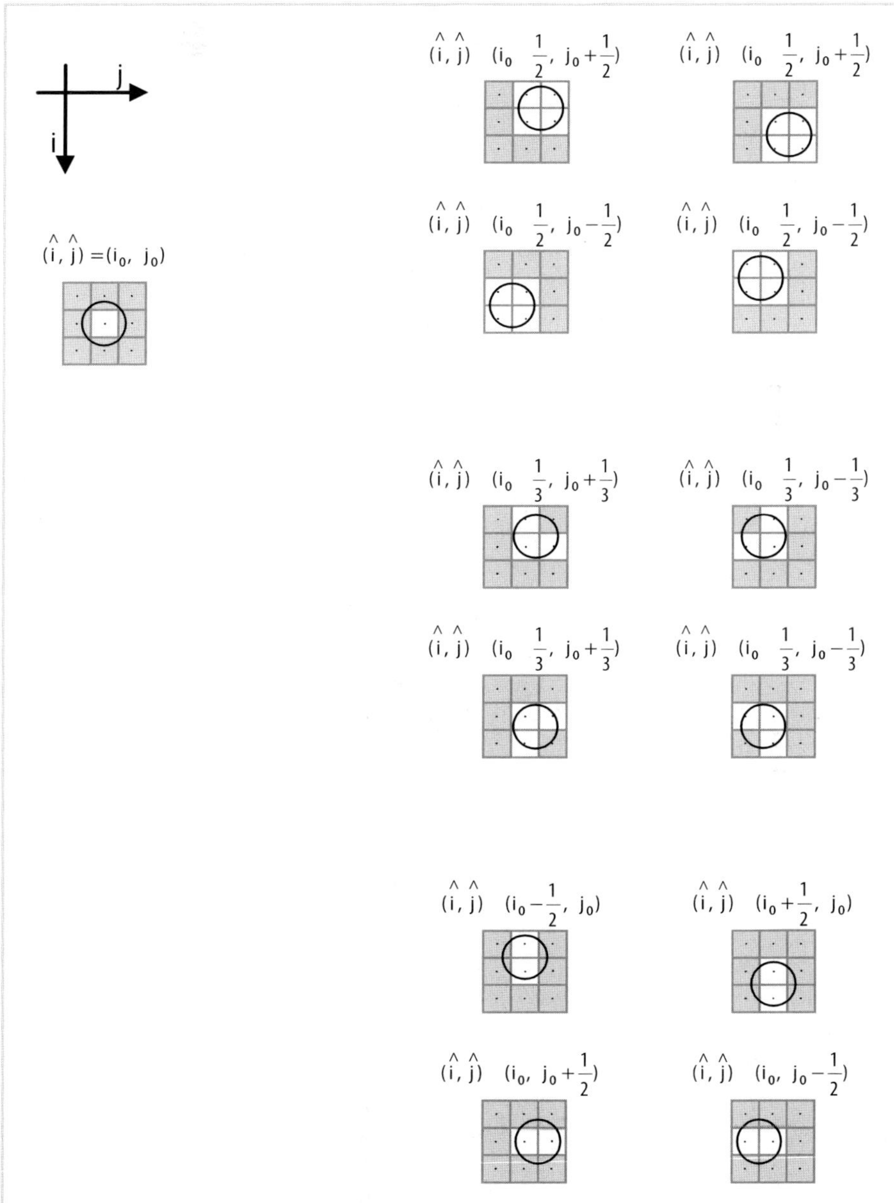

Figure 2.12. Possible digital images for a small landmark lying within $\pm 1/2$ pixel of coordinate (i_o, j_o) under the assumptions of no smoothing and 0% sensitive area. With translations over a $\pm 1/2$ pixel range, this landmark will produce one of 13 possible digital images.

Figure 2.12 shows that sweeping the location of the landmark over a range of $\pm 1/2$ pixel gives thirteen possible digital images. When the digital image is passed to a location estimation algorithm, it will give one of 13 possible results. Thus, the continuum of possible true landmark locations maps to a finite set of digital images and estimated landmark locations, and each possible digital image corresponds to an area of possible true landmark locations. This area of possible true landmark locations corresponding to one digital image is called a *locale*. Locales were first considered by Havelock (1989), who looked at questions of landmark location uncertainty when processing binary images. Notice that the example above doesn't require binary intensity quantization. By considering each pixel to have a point of sensitive area (called impulsive sampling), and by assuming that the illuminance function is discontinuous at the edge of the landmark, the pixels only take the values of A and 0, even if no intensity quantization is considered.

All landmark locations lying within $\pm 1/2$ pixel of image coordinate (i_o, j_o) will generate one of the 13 digital images shown in Figure 2.12. No matter which estimator is used, each possible digital image can give rise to only one value for estimated landmark location. Thus a locale of possible true image locations will give rise to a single value of estimated landmark location. The skew error is the distance between the true center of the landmark and the centroid given above. Within a locale, the centroid doesn't change, and so the skew error, which is the estimator bias, depends on where within the locale the landmark center falls.

Finite Sensitive Area, Pixel Geometry, and Noise Model

In practice, the analog intensity level of each pixel is the average energy level captured in its sensitive area. That is, the spatial sampling is not impulsive. The sensitive area of the pixel is the portion of the pixel surface that captures incoming photons for subsequent conversion to the electrical intensity signal. The percentage of sensitive area within a pixel boundary is called the fill factor of the pixel. Fill factors range from 25% in low-cost imagers up to 100% in specialized imagers used in astronomy, where every photon counts. Fill factors ranging between 25% and 80% are typical of most commercial digital cameras (Jahne 1997).

For mathematical convenience, the sensitive area is modeled as a rectangular-shaped region embedded within the imaginary pixel boundary grid. With finite sensitive area, the spatial sampling process can be viewed as the spatial average of the multiplication of the continuous image illuminance function, $E(x, y)$, by a pixel sampling function $p(x, y)$. The pixel sampling function is given by:

$$p(x,y) = \sum_i \sum_j \prod \left(\frac{(x - i\Delta x)}{w_x}, \frac{(y - j\Delta y)}{w_y} \right), \tag{2.25}$$

where

$$\prod \left(\frac{(x - c_x)}{w_x}, \frac{(y - c_y)}{w_y} \right) = \begin{cases} 1 & \left\{ x, y : |x - c_x| \leq \frac{w_x}{2}, |y - c_y| \leq \frac{w_y}{2} \right\} \\ 0 & \text{otherwise} \end{cases} \tag{2.26}$$

is the 2-D rectangle function centered on the coordinate (c_x, c_y) and with widths w_x and w_y in the x and y coordinate respectively. An illustration of the function $p(x, y)$ is shown in Figure 2.13.

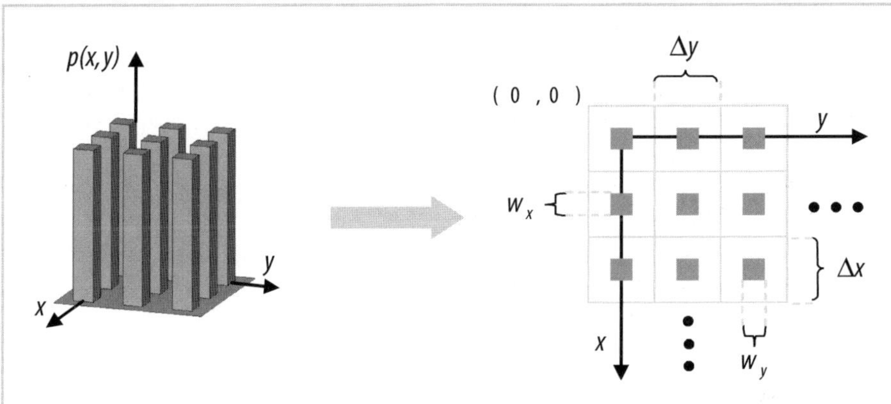

Figure 2.13. View of the pixel sampling function $p(x, y)$

The discrete analog image function will then be:

$$J(i,j) = \frac{1}{w_x w_y} \int_{i\Delta x - \frac{w_x}{2}}^{i\Delta x + \frac{w_x}{2}} \int_{j\Delta y - \frac{w_y}{2}}^{j\Delta y + \frac{w_y}{2}} E(x,y)\, p(x,y)\, dy\, dx + \widetilde{J}(i,j) , \tag{2.27}$$

where $\widetilde{J}$ is a random noise contribution. Here a Gaussian-distributed noise with zero mean and a variance of σ_n^2 is assumed. This noise model is simplified, but sufficient for the present purposes and will support development of the CRLB. Gaussian noise has been considered in the past for semiconductor imaging systems (Gonzales and Woods 1992).

For this framework, the fill factor, F_f, of the imager can be expressed as:

$$F_f = \frac{w_x\, w_y}{\Delta x\, \Delta y} . \tag{2.28}$$

The present work is the first to consider finite fill factor and landmark location precision. In Chapter 8, we analyze a configuration with sensitive area ranging from 0% (impulsive sampling) to 100% fill factor, and the impact on landmark location uncertainty is shown.

Intensity Quantization

Intensity quantization transforms the discrete analog intensity function J into the digital image function I. This transformation introduces a quantization error. In conjunction with the effects of additive Gaussian noise, the intensity quantization process adds uncertainty to landmark location. Several authors have investigated methods to minimize the effects of intensity quantization in digital image formation using statistical approximations (Dvornychenko 1983; Förstner 1982; Nielsen *et al.* 1984; Bruckstein 1987). Tian and Huhns (1986) were the first to relate the number of bits per pixel to landmark location uncertainty. Nielsen *et al.* (1984) and Bruckstein (1987) have studied intensity quantization using a statistical framework independent of landmark geometry.

2.2 Spatial and Intensity Quantization and Locales

Havelock (1989, 1991) has developed a general analytic framework for landmark location uncertainty based on the locales introduced by spatial and intensity quantization. As described above, a locale is a region on the imager within which all locations of the true landmark lead to the same digital image. This many-to-one mapping is illustrated in Figure 2.14 for four pixel fill factors and two intensity quantization levels, giving eight locale maps. The locale map of Figure 2.14 is also shown in color in Plate 2 (see Color Section, page 145).

As with Figure 2.12, a 1.8-pixel-diameter landmark and $\sigma_b = 0$ are considered to generate Figure 2.14. Using the software tools described in Chapter 6, each locale map is generated by modeling the landmark at each point of a grid of locations covering one pixel and determining the digital image produced. These digital images are then classified, and all of the grid points that give the same digital image correspond to one locale. The colors used to distinguish locales in Figure 2.14 are assigned arbitrarily.

The upper row of Figure 2.14 shows the case for an intensity quantization with $q = 1$, or binary images. In this case there are 13 locales (out of $2^9 = 512$ possible digital images, since there are nine affected pixels in Figure 2.12). For a sensitive area of 0% and a quantization of $q = 1$, the 13 digital images of Figure 2.12 correspond to the upper-right locale map of Figure 2.14. The lower row of Figure 2.14 illustrates the case for an intensity quantization of $q = 2$, or four intensity levels, in which case there are 65 locales, out of 2^{18} possible digital images.

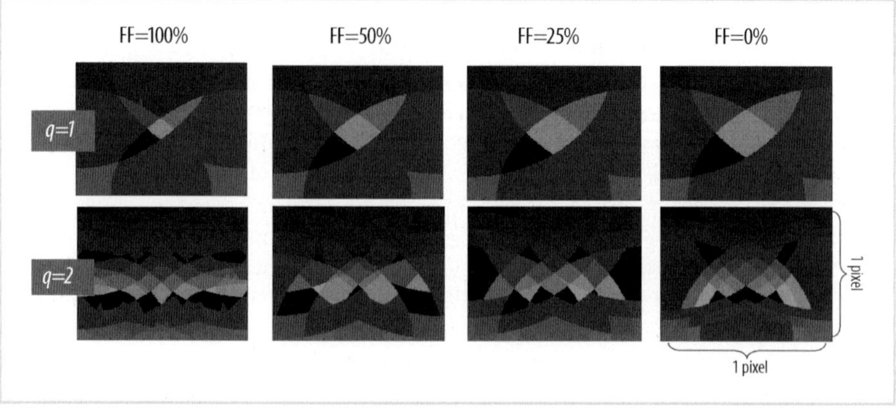

Figure 2.14. Example locale maps for four values of fill factor and to levels of intensity quantization. Each of the upper-row cases has 13 locales, corresponding to the 13 configurations shown in Figure 2.12.

Four locale maps are shown in each row, corresponding to four levels of fill factor. The changing shapes of the locales as one goes from 0% to 100% fill factor illustrates the impact of sensitive area on the detection of landmark location.

The size and shape of each locale corresponds to the distance the landmark must be displaced to change the digital image. As such, locale dimensions are indicative of the local derivative of the analog intensity function with respect to location, $(\partial J/\partial\, {}^{p}P_{o}^{*})$. A remarkable characteristic of locales is their irregularity, indicating the nonstationary statistics of landmark location. In some regions of the image, a small shift of the landmark produces a change in the digital image, while in other regions a large shift may not change the digital image.

The irregularity of the locales is also seen in the range of their aspect ratios; that is, the ratios of the longest to the shortest dimensions. For example, in the lower-right locale map, some locales are approximately isotropic, which is to say they have an aspect ratio of approximately one, while others have a high aspect ratio. The variability of locale aspect ratio also shows the variability of $(\partial J/\partial\, {}^{p}P_{o}^{*})$ as the true landmark location changes.

Practical situations, of course, involve larger landmarks and higher resolution intensity quantization. Nonetheless, the essential features of locales carry over. The interaction of imager noise, $\widetilde{J}$, and locales is considered in Section 3.2, where suitable analysis tools are developed.

With a larger landmark and more bits of intensity resolution there are more locals, but irregularity remains a hallmark. For example, a study of a circular landmark with a diameter of three pixels and a quantization of $q = 4$ gives 10,557 locales. While the average area is $1/10{,}557 \sim 10^{-4}$ pixels^2, corresponding to hypothetical linear dimensions of 10^{-2} pixels, the actual largest locale has an area of 0.0014 pixels^2 and linear dimensions of 0.052 by 0.003 pixels, while the smallest has an area of 1.5×10^{-11} pixels^2 and linear dimensions of 9.1×10^{-6} by 4.3×10^{-6} pixels, or approximately 5×10^3 smaller in linear dimensions than the largest locale. These two locales have relatively low aspect ratios. Another locale of the study case has an aspect ratio of 8,000, measuring approximately 3.4×10^{-2} by 4.3×10^{-6} pixels.

The properties of locales motivate two aspects of the treatment of landmark location uncertainty. The first is consideration of the true landmark location, ${}^{p}P_{o}^{*}$, as a uniformly distributed random variable, distributed over an interval of $\pm 1/2$ pixel. The second is use of the 95% (or 99%) confidence interval to express uncertainty.

Considering locales, it is seen that the statistics of landmark location estimation depend on precisely where the landmark falls on the imager. Since there is generally no way to know precisely where the landmark center has fallen, the practical approach to uncertainty is to treat the true landmark location, ${}^{p}P_{o}^{*}$, as a uniformly distributed random variable. Because the statistics repeat with a shift of an integer number of pixels, it is sufficient to study the statistics arising with a uniform distribution on an area of one pixel, or from $-1/2$ to $+1/2$ pixel on each axis. The $\pm 1/2$ pixel region, example locales, and repetition of the locales with a one-pixel shift are illustrated in Figure 2.15.

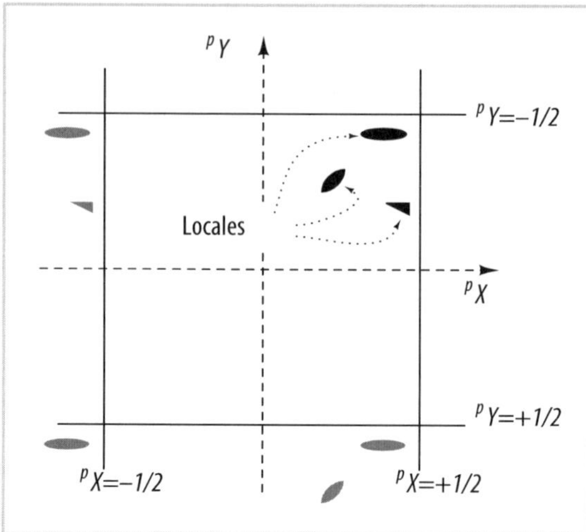

Figure 2.15. Illustration of locales and $\pm 1/2$ pixel area of uniform probability distribution for ${}^pP_o^*$

A second aspect of the treatment of landmark location uncertainty is the proper consideration of estimator bias and covariance. Bias is defined as

$$ {}^pP_0^{\text{bias}} = {}^pP_o^* - {}^p\widehat{P}_0^* , \tag{2.29} $$

where ${}^pP_o^*$ is the true landmark location, and

$$ {}^p\widehat{P}_0^* = T\left(I^*\right) \tag{2.30} $$

is the estimated location given by applying estimator $T(\cdot)$ to the noise-free digital image, $I^*\left({}^pP_o^*\right)$. Actual imagers have noise, so that the measured digital image, $\bar{I}$, is a random variable. Treatment of the statistics of $\bar{I}$ follow in Section 3.2.2. Here, the important idea is that the ensemble of $\bar{I}$ with non-negligible probability gives a distribution of actual landmark location estimates that approximates ${}^p\widehat{P}_0^*$ in the mean and has a covariance given by

$$ \Sigma = \text{Cov}\,{}^p\widehat{P}_0 = E\left\{\left({}^p\widehat{P}_0 - {}^p\widehat{P}_0^*\right)^T\left({}^p\widehat{P}_0 - {}^p\widehat{P}_0^*\right)\right\} . \tag{2.31} $$

Approximating the distribution of ${}^p\widehat{P}_0$ with a 2-D Gaussian distribution, the combination of bias and covariance gives rise to a probability distribution that is offset from ${}^pP_o^*$ and distributed over an elliptical region. This is illustrated in Figure 2.16, where the ellipse is drawn according to:

$$ \left({}^p\widehat{P}_0 - {}^p\widehat{P}_0^*\right)^T \Sigma^{-1}\left({}^p\widehat{P}_0 - {}^p\widehat{P}_0^*\right) = c , \tag{2.32} $$

where c is a constant that is related by the χ^2 distribution to the number of samples expected to lie inside the ellipse.

The bias and covariance matrix can be computed for a specified configuration and ${}^pP_o^*$, but this is not directly useful, since these values vary rapidly over the

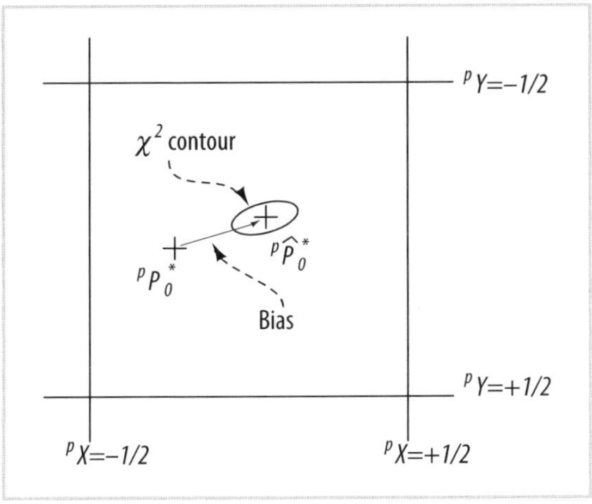

Figure 2.16. Illustration of estimator covariance and bias

±1/2 pixel area. In fact, when the landmark has two axes of symmetry, the average bias is zero, as seen be the absence of a preferred direction. A method for representing the uncertainty for this system with nonstationary statistics is required. The method chosen for this study is the circular 95% confidence interval. It is the radius of a circle centered on $^PP_o^*$ in which 95% of the estimates $^P\widehat{P}_0$ will fall, considering, for each true landmark location $^PP_o^*$, the bias and covariance illustrated in Figure 2.16, and also considering the uniform distribution of $^PP_o^*$ over the area of one pixel. Computation of the confidence interval will require a four-dimensional integral. For each point $^PP_o^*$, a 2-D integral over the Gaussian distribution of $^P\widehat{P}_0$ arising from $^PP_o^*$; and then a 2-D integral over uniform distribution of $^PP_o^*$ over the area of one pixel. Details of these integrals are the subject of Chapter 3.

2.3 Illustrative Example

We conclude this chapter with an illustrative example. On the one hand, the example shows the use of a confidence interval and performance map; and on the other, the impacts of sensitive area and diffraction smoothing are illustrated. The analysis was performed using a synthetic image of a landmark by starting with the landmark geometrical description and applying the image formation transformations indicated in Equation 2.4. For convenience, Figure 2.17 repeats these transformations, highlighting the user configuration parameters used.

Plate 3 (see Color Section, page 146) shows the distribution of

$$^P\widetilde{P}_0 = {^P\widehat{P}_0} - {^PP_o^*} \tag{2.33}$$

for twelve cases, corresponding to three levels of fill factor, $F_f = \{0\%, 50\%, 100\%\}$ (shown with color in each subfigure), and to two values of smoothing parameter, $\sigma_b = \left\{10^{-3}, 5 \times 10^{-3}\right\}$ (in millimeters) on the imager, as well as to two values

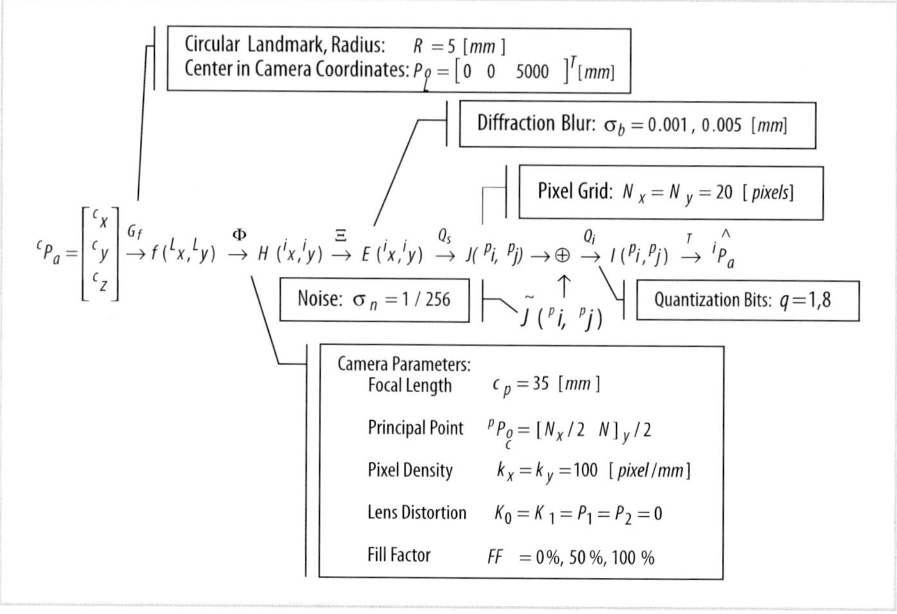

Figure 2.17. Configuration parameters for case study

of quantization resolution, $q = \{1, 8\}$. To visualize this figure, imagine that ${}^{p}P_o^*$ is the center for each of 100 samples. The points plotted then correspond to ${}^{p}\widetilde{P}_o$, and the confidence interval is the radius of the circle enclosing 95% of the samples. The samples of ${}^{p}P_o^*$ are distributed randomly across the interval of $\pm 1/2$ pixels. Landmark location ${}^{p}\widehat{P}_0$ has been estimated using the centroid estimator of Equation 2.23 and Equation 2.24. This case study is made using the tools presented in Chapter 6.

Plate 3 shows that use of grayscale information in the centroid algorithm can reduce uncertainty by 25% (we will see in Chapters 7 and 8 that model-based estimators can give a sixfold improvement). Plate 3 also shows that an enlarged sensitive area can reduce uncertainty, particularly for a sharply focused image ($\sigma_b = 0.001$ mm). Finally, Plate 3 shows that increasing the image smoothing, for example by defocusing the image, can improve landmark location precision.

In the next chapter the model developed in Section 2.1 and the methodology outlined in Section 2.2 are applied to determine the Cramér–Rao lower bound and also analysis tools for determining the uncertainty performance of practical estimators.

3 Analytic Framework for Landmark Location Uncertainty

In this chapter the mathematical tools for analyzing landmark location uncertainty are developed. Two cases are considered. In Section 3.1 tools for calculating the Cramér–Rao lower bound are developed, and in Section 3.2 tools are developed for analyzing practical estimators. The two cases are quite different. The Cramér–Rao lower bound assumes an unbiased estimator, and so the uncertainty is entirely dependent on the covariance component, determined by $\left(\partial J/\partial\, {}^{p}P_{o}^{*}\right)$. Practical estimators have bias that is dependent on the true landmark location, ${}^{p}P_{o}^{*}$. Interesting challenges arise when performing the integration required, because the bias and covariance can change so rapidly with ${}^{p}P_{o}^{*}$.

3.1 Cramér–Rao Lower Bound

The Cramér–Rao lower bound methodology provides a theoretical statistical minimum limit on the landmark location uncertainty and institutes a general framework for its study (Van Trees 1968). The ability to place a lower bound on landmark location uncertainty provides a benchmark for comparisons of the performance of practical estimators and offers guidelines for the design of new estimators with improved performance (Kay 1993).

The Cramér–Rao inequality has been widely used in statistical signal processing applications. Some examples are the work of Reza and Doroodchi (1996), who established the CRLB for the problem of detecting sudden changes in step-like signals, Armstrong and Holeman (1998) in the determination of motion parameter uncertainty for Doppler radar tracking, and numerous efforts in the field of time delay estimation in communication systems and sonar (Scharf 1990; Scharf and McWhorter 1992). Applications of the CRLB in the field of image processing include the work from Hung (1991), who determined the CRLB for the image restoration problem, and also the framework developed by Robinson and Milanfar (2004), who established the CRLB on the problem of interframe motion estimation in digital images.

A barrier to the development of the CRLB for landmark location uncertainty is the difficulty involved in determining the spatial partial derivatives required, $\left(\partial J/\partial\, {}^{p}P_{o}^{*}\right)$. There are several obstacles to computing the spatial partial derivatives. First, digital images are discontinuous functions of landmark location due to amplitude quantization. Second, when smoothing is considered, the illuminance function is not analytically tractable, making closed-form mathematical formulation of the spatial partial derivatives impossible and necessitating the use of numerical integration. The analog intensity function J and its derivatives must be computed by numerical integration of the illuminance function, $E({}^i x, {}^i y)$, which is performed by evaluating $E({}^i x, {}^i y)$ on a sample grid with a resolution of many points per pixel. Because the illuminance function is given by numerical convolution of the optical flux function with the smoothing operator, it is expensive to compute; and integration of the confidence interval contributions over the $\pm 1/2$ pixel distribution of ${}^p P_o^*$ requires that the integrations be performed many times, each with a slightly shifted landmark location. Previously known methods of 2-D numerical integration require the sample grid to be collocated with the region of integration (for details, see Figure 5.1). The use of an available method for 2-D numerical integration would require that the convolution be recomputed on a new sample grid for each position of the landmark evaluated, which would lead to prohibitively long computation times.

A second challenge is the nonstationarity of the statistics of $\widehat{{}^p P_0}$, or, equivalently, the strong dependence of $\left(\partial J / \partial\, {}^p P_o^*\right)$ on the precise value of ${}^p P_o^*$. Addressing this challenge, several authors make use of the expected value of the CRLB over the event space, creating a single-valued bound (*e.g.*, Reza and Doroodchi 1996). Taking a different path, we use the confidence interval to characterize landmark location uncertainty.

Considering these challenges, the strategy of this research comprises:

- Extensive modeling of image formation, including smoothing, which enables the numerical determination of the required spatial partial derivatives.
- Development of statistical confidence intervals based on the Cramér–Rao lower bound. The bound will be calculated on the discrete analog amplitude function J.
- Development of a method of 2-D numerical integration that can be applied when the region of integration is not collocated with the sample grid.

These developments are described in Sections 3.1.1 and 3.1.2, and the integration method is described in detail in Chapter 5.

3.1.1 Analytic Framework for Cramér–Rao Lower Bound

The derivation of the landmark location CRLB on the discrete analog intensity function J follows. The signal model for this problem is:

$$\bar{J}\left(m, n, {}^p P_o^*\right) = J^*\left(m, n, {}^p P_o^*\right) + \widetilde{J}(m, n) \tag{3.1}$$

where $\bar{J}\left(m, n, {}^p P_o^*\right)$ is the discrete analog intensity function prior to quantization, $J^*\left(m, n, {}^p P_o^*\right)$ is the discrete analog intensity function, ideally produced by the landmark at location ${}^p P_o^*$, and $\widetilde{J}(m, n)$ is a zero-mean white stationary Gaussian process with variance, σ_n^2, and where m and n are indices on the pixel array.

The joint likelihood function for the data $\bar{J}(m,n;{}^{p}P_{o}^{*})$; $n = 0,1,2,\ldots,N-1$, $m = 0,1,2,\ldots,M-1$; is related to the probability density function of $\widetilde{J}(m,n)$, and is given by:

$$p(\bar{J};{}^{p}P_{o}^{*}) = \prod_{m,n} \frac{1}{\sqrt{2\pi\sigma_n^2}}\, \mathrm{e}^{-\frac{(\bar{J}(m,n,{}^{p}P_{o}^{*})-J^{*}(m,n;{}^{p}P_{o}^{*}))^2}{2\sigma_n^2}}$$

$$= \left(2\pi\sigma_n^2\right)^{-\frac{mn}{2}} \mathrm{e}^{-\frac{1}{2\sigma_n^2}\sum\limits_{m,n}\left(\bar{J}(m,n,{}^{p}P_{o}^{*})-J^{*}(m,n;{}^{p}P_{o}^{*})\right)^2} . \qquad (3.2)$$

The logarithm of the likelihood function is:

$$\log p(\bar{J};{}^{p}P_{o}^{*}) = -\frac{mn}{2}\log\left(2\pi\sigma_n^2\right) - \frac{1}{2\sigma_n^2}\sum_{m,n}\left(\bar{J}(m,n;{}^{p}P_{o}^{*}) - J^{*}(m,n;{}^{p}P_{o}^{*})\right)^2 . \qquad (3.3)$$

The Fisher information matrix (FIM) is given by:

$$\left[J\left({}^{p}P_{o}^{*}\right)\right]_{i,j} = -E\left\{\frac{\partial^2 \log p(\bar{J};{}^{p}P_{o}^{*})}{\partial\, {}^{p}P_{i}^{*}\,\partial\, {}^{p}P_{j}^{*}}\right\}, \qquad (3.4)$$

where $E\{\bullet\}$ is the expected value operator, $i,j \in \{1,2\}$ are indices on the two axes in image coordinates, ${}^{p}P_{0}^{*} = \left[{}^{p}P_{1}^{*}\ {}^{p}P_{2}^{*}\right]^{T}$ is the true landmark location, and $\left[J\left({}^{p}P_{o}^{*}\right)\right]_{i,j}$ is the (i,j) element of the FIM. To determine the argument of the FIM in Equation 3.4, we proceed to establish the spatial partial derivative of Equation 3.3:

$$\frac{\partial \log p(\bar{J};{}^{p}P_{o}^{*})}{\partial\, {}^{p}P_{i}^{*}} = \frac{1}{\sigma_n^2}\sum_{m,n}\left[\left(\bar{J}(m,n;{}^{p}P_{o}^{*}) - J^{*}(m,n;{}^{p}P_{o}^{*})\right)\frac{\partial J^{*}(m,n;{}^{p}P_{o}^{*})}{\partial\, {}^{p}P_{i}^{*}}\right]. \qquad (3.5)$$

From Equation 3.5, we get:

$$\frac{\partial^2 \log p(\bar{J};{}^{p}P_{o}^{*})}{\partial\, {}^{p}P_{i}^{*}\,\partial\, {}^{p}P_{j}^{*}} = \frac{\partial}{\partial\, {}^{p}P_{j}^{*}}\left[\frac{\partial \log p(\bar{J};{}^{p}P_{o}^{*})}{\partial\, {}^{p}P_{i}^{*}}\right]$$

$$= \frac{1}{\sigma_n^2}\frac{\partial}{\partial\, {}^{p}P_{j}^{*}}\left[\sum_{m,n}\left[\left(\bar{J}(m,n;{}^{p}P_{o}^{*}) - J^{*}(m,n;{}^{p}P_{o}^{*})\right)\frac{\partial J^{*}(m,n;{}^{p}P_{o}^{*})}{\partial\, {}^{p}P_{i}^{*}}\right]\right]$$

$$= \frac{1}{\sigma_n^2}\left[\sum_{m,n}\left[\left(\bar{J}(m,n;{}^{p}P_{o}^{*}) - J^{*}(m,n;{}^{p}P_{o}^{*})\right)\frac{\partial^2 J^{*}(m,n;{}^{p}P_{o}^{*})}{\partial\, {}^{p}P_{i}^{*}\,\partial\, {}^{p}P_{j}^{*}}\right.\right.$$

$$\left.\left. - \frac{\partial J^{*}(m,n;{}^{p}P_{o}^{*})}{\partial\, {}^{p}P_{i}^{*}}\frac{\partial J^{*}(m,n;{}^{p}P_{o}^{*})}{\partial\, {}^{p}P_{j}^{*}}\right]\right]. \qquad (3.6)$$

Substituting Equation 3.6 into Equation 3.4:

$$\left[J\left({}^{p}P_{o}^{*}\right)\right]_{1,1} = -E\left\{\frac{\partial^2 \log p(\bar{J};{}^{p}P_{o}^{*})}{\partial\, {}^{p}P_{1}^{*2}}\right\} = \frac{1}{\sigma_n^2}\sum_{m,n}\left(\frac{\partial J^{*}(m,n;{}^{p}P_{o}^{*})}{\partial\, {}^{p}P_{1}^{*}}\right)^2 \qquad (3.7)$$

$$\left[J\left({}^{p}P_{o}^{*}\right)\right]_{2,2} = -E\left\{\frac{\partial^{2}\log p(\bar{J};{}^{p}P_{o}^{*})}{\partial\, {}^{p}P_{2}^{*2}}\right\} = \frac{1}{\sigma_{n}^{2}}\sum_{m,n}\left(\frac{\partial J^{*}(m,n;{}^{p}P_{o}^{*})}{\partial\, {}^{p}P_{2}^{*}}\right)^{2} \tag{3.8}$$

$$\begin{aligned}\left[J\left({}^{p}P_{o}^{*}\right)\right]_{2,1} &= \left[J\left({}^{p}P_{o}^{*}\right)\right]_{1,2}\\ &= -E\left\{\frac{\partial^{2}\log p(\bar{J};{}^{p}P_{o}^{*})}{\partial\, {}^{p}P_{1}^{*}\partial\, {}^{p}P_{2}^{*}}\right\} = \frac{1}{\sigma_{n}^{2}}\sum_{m,n}\frac{\partial J^{*}(m,n;{}^{p}P_{o}^{*})}{\partial\, {}^{p}P_{1}^{*}}\frac{\partial J^{*}(m,n;{}^{p}P_{o}^{*})}{\partial\, {}^{p}P_{2}^{*}}\,. \end{aligned} \tag{3.9}$$

Using the chain rule:

$$\frac{\partial J^{*}(m,n;{}^{p}P_{o}^{*})}{\partial\, {}^{p}P_{1}^{*}} = \frac{\partial J^{*}(m,n;{}^{p}P_{o}^{*})}{\partial x}\frac{\partial x}{\partial\, {}^{p}P_{1}^{*}} = \frac{\partial J^{*}(m,n;{}^{p}P_{o}^{*})}{\partial x} \tag{3.10}$$

$$\frac{\partial J^{*}(m,n;{}^{p}P_{o}^{*})}{\partial\, {}^{p}P_{2}^{*}} = \frac{\partial J^{*}(m,n;{}^{p}P_{o}^{*})}{\partial y}\frac{\partial y}{\partial\, {}^{p}P_{2}^{*}} = \frac{\partial J^{*}(m,n;{}^{p}P_{o}^{*})}{\partial y}\,. \tag{3.11}$$

Substituting Equation 3.10 and Equation 3.11 into Equation 3.7 through Equation 3.9, we obtain the Fisher information matrix:

$$J\left({}^{p}P_{o}^{*}\right) = \frac{1}{\sigma_{n}^{2}}\sum_{m,n}\begin{bmatrix}\left(\dfrac{\partial J^{*}(m,n;{}^{p}P_{o}^{*})}{\partial x}\right)^{2} & \dfrac{\partial J^{*}(m,n;{}^{p}P_{o}^{*})}{\partial x}\dfrac{\partial J^{*}(m,n;{}^{p}P_{o}^{*})}{\partial y}\\ \dfrac{\partial J^{*}(m,n;{}^{p}P_{o}^{*})}{\partial x}\dfrac{\partial J^{*}(m,n;{}^{p}P_{o}^{*})}{\partial y} & \left(\dfrac{\partial J^{*}(m,n;{}^{p}P_{o}^{*})}{\partial y}\right)^{2}\end{bmatrix}. \tag{3.12}$$

Finally, the Cramér–Rao lower bound on the variance of the estimate of the center of the landmark, ${}^{p}\widehat{P}_{o}$, is established (Kay 1993):

$$\begin{aligned}E\left\{\left({}^{p}\widehat{P}_{o} - {}^{p}P_{o}^{*}\right)\left({}^{p}\widehat{P}_{o} - {}^{p}P_{o}^{*}\right)^{T}\right\} &\geq \frac{\partial E\left\{{}^{p}\widehat{P}_{o}\right\}}{\partial\, {}^{p}P_{o}^{*}} J\left({}^{p}P_{o}^{*}\right)^{-1}\frac{\partial E\left\{{}^{p}\widehat{P}_{o}\right\}^{T}}{\partial\, {}^{p}P_{o}^{*}}\\ &\quad + \left(E\left\{{}^{p}\widehat{P}_{o}\right\} - {}^{p}P_{o}^{*}\right)\left(E\left\{{}^{p}\widehat{P}_{o}\right\} - {}^{p}P_{o}^{*}\right)^{T}\,.\end{aligned} \tag{3.13}$$

For some estimation problems, biased estimators are known that can give lower covariance than unbiased estimators. However, useful biased estimators are not known for landmark location, and are not likely because the bias changes rapidly as ${}^{p}P_{o}^{*}$ changes. Considering only unbiased estimators, then $E\left\{{}^{p}\widehat{P}_{o}\right\} - {}^{p}P_{o}^{*} = 0$ and $\frac{\partial E\{{}^{p}\widehat{P}_{o}\}}{\partial\, {}^{p}P_{o}^{*}} = I$, where I is the identity matrix. Using these the CRLB expression is simplified to:

$$E\left\{\left({}^{p}\widehat{P}_{o} - {}^{p}P_{o}^{*}\right)\left({}^{p}\widehat{P}_{o} - {}^{p}P_{o}^{*}\right)^{T}\right\} \geq J\left({}^{p}P_{o}^{*}\right)^{-1}\,, \tag{3.14}$$

which gives the CRLB covariance as:

$$\Sigma_{\mathrm{CRLB}} = J\left({}^{p}P_{o}^{*}\right)^{-1} \tag{3.15}$$

With a landmark at ${}^{p}P_{o}^{*}$, Equation 3.15 gives the minimum covariance that any unbiased estimator can achieve. In the next section, this covariance is used in conjunction with a uniform distribution of ${}^{p}P_{o}^{*}$ across $\pm 1/2$ pixel, to determine the smallest confidence interval possible for unbiased landmark location estimation.

3.1.2 CRLB Confidence Interval

Figure 3.1 illustrates the CRLB covariance uncertainty contours at nine positions inside the ±0.1 pixel area of a hypothetical configuration. The nine ellipses of Figure 3.1 represent the 95% confidence intervals for the covariance of a given estimate ${}^{P}\widehat{P}_o$ (shown with a dot) of a true center ${}^{P}P_o^*$ (shown with a cross). A 4-D integration is required to form the confidence interval. At each point ${}^{P}P_o^*$, a 2-D integration of the Gaussian-distributed probability density function for ${}^{P}\widehat{P}_o$ is required to get the confidence interval contribution at ${}^{P}P_o^*$ (Equation 3.16, below), and then a 2-D integration across the uniformly distributed probability density function for ${}^{P}P_o^*$ (Equation 3.20, below).

The CRLB cumulative probability function for an estimate ${}^{P}\widehat{P}_o$, at a distance R from ${}^{P}P_o^*$ is defined as:

$$F_{{}^{P}\widehat{P}_o}^{\mathrm{CRLB}}\left\{R;{}^{P}P_o^*\right\} = p\left(\left\|{}^{P}\widehat{P}_o - {}^{P}P_o^*\right\| < R\right) = \int_{r=0}^{R}\int_{\alpha=0}^{2\pi} rG({}^{P}P;{}^{P}P_o^*,\Sigma_{\mathrm{CRLB}})\,d\alpha\,dr\,, \tag{3.16}$$

where $G\left({}^{P}P;{}^{P}P_o^*,\Sigma_{\mathrm{CRLB}}\right)$ is a Gaussian probability function centered on ${}^{P}P_o^*$ and covariance Σ_{CRLB}:

$$G\left({}^{P}P,{}^{P}P_o^*\right) = \frac{1}{2\pi\sqrt{|\Sigma_{\mathrm{CRLB}}|}}\,\mathrm{e}^{-\frac{1}{2}\left({}^{P}P-{}^{P}P_o^*\right)^T\Sigma_{\mathrm{CRLB}}^{-1}\left({}^{P}P-{}^{P}P_o^*\right)}\,, \tag{3.17}$$

where $|\Sigma_{\mathrm{CRLB}}|$ is the determinant of the CRLB covariance matrix.

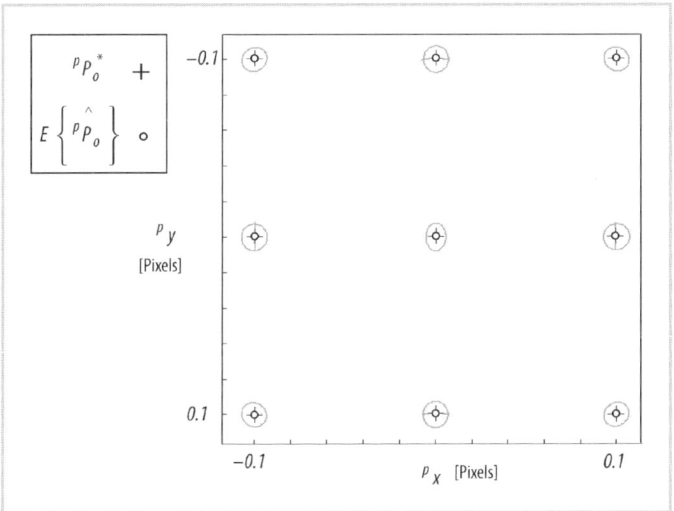

Figure 3.1. CRLB uncertainty contours at different position estimates

Consequently, the CRLB radial probability density function for a given position estimate of ${}^{p}\widehat{P}_o$ is defined as:

$$f^{\mathrm{CRLB}}_{{}^{p}\widehat{P}_o}\left(R;{}^{p}P^*_o\right) = \frac{dF^{\mathrm{CRLB}}_{{}^{p}\widehat{P}_o}\left\{R;{}^{p}P^*_o\right\}}{dR}\,, \tag{3.18}$$

and correspondingly the CRLB confidence interval for this estimate is:

$$C^{\mathrm{CRLB}}_n\left({}^{p}P^*_o\right) = \min\left\{R \,\Big|\, \int_{r=0}^{R} f^{\mathrm{CRLB}}_{{}^{p}\widehat{P}_o}\left(r;{}^{p}P^*_o\right)\,dr = n\right\}\,, \quad 0 < n < 1\,, \tag{3.19}$$

where n is the confidence level, for example $n = 0.95$. Finally, the overall confidence interval is created by the integration of Equation 3.19 in the $\pm 1/2$ pixel region Ω:

$$\overline{C_n}^{\mathrm{CRLB}} = \iint_{\Omega} C^{\mathrm{CRLB}}_n\left({}^{p}P^*_o\right)\,dx\,dy\,. \tag{3.20}$$

The numerical integration of Equation 3.20 is carried out using the 2-D Simpson's rule. In Chapter 8, results are presented where the 4-D integrations of Equations 3.16 and 3.20 are evaluated numerically to determine the Cramér–Rao lower bound on uncertainty for a range of configurations.

3.2 Uncertainty of Practical Estimators

In the previous section, methods were developed to compute a lower bound on landmark uncertainty for any estimator. In this section, the performance of practical estimators is considered, which will bring two new challenges. First, both image noise and intensity quantization must be considered. As before, for a given ${}^{p}P^*_o$, a set of ideal discrete analog intensities are computed, J^*. These are quantized to give the ideal digital image, $I^* = Q_i(J^*)$. The ideal digital image is also one possible measured image, corresponding to $\overline{I} = I^*$. Because of image noise, the measured image is a random variable, with each event corresponding to one possible value of $\overline{I} = Q_i(J^* + \widetilde{J})$. The operation of the quantizer transforms the effect of image noise from a continuous probability density function to a set of discrete events, each with finite probability. Assuming that the samples of image noise in each pixel are independent and Gaussian-distributed, for each pixel j there may be only a few values of $\overline{I}(j)$ with non-negligible probability. The set of events for landmark location, however, is the set of digital images $\overline{I}$ with non-negligible probability, which is the power set of the events for each pixel. The number of possible events $\overline{I}$ can be a very large number indeed. In Section 3.2.1, a method is developed to compute an approximation to the covariance matrix of an estimator ${}^{p}\widehat{P}_0 = T(\overline{I})$ corresponding to the set of events $\overline{I}$ and a specific choice of estimator, without considering each of the many events individually.

The second new challenge when considering practical estimators is the rapid variation of bias and covariance as ${}^{p}P^*_o$ changes; that is, the nonstationary statistics of $\widehat{{}^{P}P_0}$. Two-dimensional numerical integration operates by fitting a polynomial surface to values of the integrand and forming the definite integral of the polynomial surface. The correctness of this approach depends on the integrand being sufficiently smooth, which is not the case for the bias and covariance of landmark location estimators. To avoid aliasing and other errors that might arise with an interaction between a regular sampling grid and the underlying and unknown structure of locales, Monte Carlo integration is used for the 2-D integration of the uniform distribution of ${}^{P}P_o^*$ over $\pm 1/2$ pixels (McCluer 2000). Monte Carlo integration has rather slow convergence, and so the numerically computed error distributions are somewhat rough for practical computing times, but the method avoids systematic errors. Recall that a 4-D integration is required to compute the distribution of $\widetilde{{}^{P}P_o}$, a 2-D integration over the biased, Gaussian ellipse corresponding to a specific ${}^{P}P_o^*$, and then a 2-D integration is necessary over the uniform distribution of ${}^{P}P_o^*$ across $\pm 1/2$ pixel. The integrand of the first 2-D integration is smooth and can be integrated by standard techniques, in this case the 2-D Simpson's rule. The latter is addressed by Monte Carlo integration. These methods are derived in Section 3.2.2.

3.2.1 Analytic Framework for Practical Estimators

In the absence of noise, the digital image is created from the amplitude quantization of the discrete analog intensity function:

$$I^* = Q_i(J^*) \,, \tag{3.21}$$

where J^* is a column vector of N_p analog intensity values corresponding to the N_p pixels comprising the image of the landmark. In the presence of noise, the digital image is expressed as:

$$\bar{I} = Q_i(J^* + \widetilde{J}) = Q_i(\bar{J}) \,, \tag{3.22}$$

where $\widetilde{J}$ is Gaussian noise with variance σ_n^2. There is a finite set of digital images, and in principle the covariance of $\widetilde{\theta}$ can be determined by forming $\widetilde{\theta}$ and the probability associated with each one. However, with even a moderate level of imager noise $\widetilde{J}$, there are an enormous number of possible digital images of the landmark, and so an efficient method of computing the covariance matrix is needed. This is done by Taylor series expansion of the covariance matrix itself.

For each pixel j, there is a finite set of events $\bar{I}(j)$ induced by the image noise $\widetilde{J}$ operating through the quantizer. The event i of pixel j is indexed by i_j, so $\bar{I}(j, i_j)$ is a specific possible quantized value for pixel j. By restricting our attention to events with probabilities that are greater than some threshold, such as $p(\bar{I}(j, i_j)) > 10^{-6}$, a number $M^{(j)}$ of events for pixel j need to be considered. For example, considering a hypothetical pixel j with eight-bit quantization, an ideal analog intensity value of $J^*(j) = 121.3$ (in ADC counts) and a noise figure of $\sigma_n = 0.6$ (in counts). Then the noise-free quantized intensity value is $I^* = 121$, and the values i_j and probabilities for several possible events of pixel j are shown in Table 3.1. Notice that the event index i_j is arranged such that $\bar{I}(j, i_j) = I^*(j) + i_j$.

Table 3.1. Example data showing enumeration of events for pixel j with $J^* = 121.3$ and $\sigma_n = 0.6$. Events with $p\left(\bar{I}(j, i_j)\right) > 10^{-6}$ are shown. $M^{(j)} = 6$ for the example pixel.

Range of analog intensity	Event index i_j	$\bar{I}(j, i_j)$	$p\left(\bar{I}(j, i_j)\right)$
$118.5 < \bar{J} \leq 119.5$	−2	119	0.001348
$119.5 < \bar{J} \leq 120.5$	−1	120	0.089861
$120.5 < \bar{J} \leq 121.5$	0	121	0.539347
$121.5 < \bar{J} \leq 122.5$	1	122	0.346671
$122.5 < \bar{J} \leq 123.5$	2	123	0.022627
$123.5 < \bar{J} \leq 124.5$	3	124	0.000123

Let M denote the total number of digital images, $\bar{I}_k$, $k = 1, \ldots, M$, that can be formed by the joint event $\{i_1, i_2, i_3, \ldots, i_N\}$; that is, the total set of images is the power set of possible pixel intensity values:

$$M = \prod_{j=1}^{N_p} M^{(j)} \tag{3.23}$$

Index k is then an index on the power set. Each k is associated with a joint event with particular values for each of $\{i_1, i_2, i_3, \ldots, i_N\}$.

From here it is possible to define the covariance of the landmark location estimator as:

$$\mathrm{Cov}\,\widetilde{\theta} = E\left\{\widetilde{\theta}\widetilde{\theta}^T\right\} = \sum_{i_1}\sum_{i_2}\sum_{i_3}\cdots\sum_{i_N} P_{i_1}P_{i_2}P_{i_3}\cdots P_{i_N}\left(\widetilde{\theta}_k\widetilde{\theta}_k^T\right), \tag{3.24}$$

where each summation is over the $M^{(j)}$ events of pixel j, P_{i_j} is a shorthand notation for the probability $p(\bar{I}(j, i_j))$, and the misadjustment of the landmark location estimate $\widetilde{\theta}_k$, corresponding to $\bar{I}_k$, is given by:

$$\begin{aligned}\widetilde{\theta}_k = \widetilde{\theta}_{\{i_1,i_2,i_3,\ldots,i_N\}} &= \widehat{\theta}_{I^*} - \widehat{\theta}_k \\ &= T(I^*) - T(\bar{I}_k)\,,\end{aligned} \tag{3.25}$$

where $T(I)$ is the output of a landmark location estimator applied to image I. Equation 3.24 is not usable in practice given the large power set M. For example, a 10×10 image where each pixel has six possible outcomes will generate 6^{100} digital images!

Rather than enumerating the large power set of possible digital images, the covariance of $\widetilde{\theta}$ is formed using the Taylor series expansion. A particular digital image $\bar{I}_k$ (caused by the joint event $\{i_1, i_2, i_3, \ldots, i_N\}$) can be expressed as:

$$\bar{I}_k = I^* + \partial I^k\,, \tag{3.26}$$

where

$$\delta I^k = \left[i_1 i_2 i_3 \cdots i_N\right]^T. \tag{3.27}$$

The estimator output is given by the Taylor series expansion:

$$T\left(\bar{I}_k\right) = \widehat{\theta}_k = \widehat{\theta}_{\{i_1,i_2,i_3,\ldots,i_N\}} = \widehat{\theta}_{I^*} + \left.\frac{\partial T}{\partial I}\right|_{I=I^*} \delta I^k + O\left(\frac{1}{2}\frac{\partial^2 T}{\partial I_{j_1}\partial I_{j_2}}\delta I_{j_1}\delta I_{j_2}\right) \tag{3.28}$$

$$= \widehat{\theta}_{I^*} + \sum_{j=1}^{N_p} \left.\frac{\partial T}{\partial I_j}\right|_{I=I^*} \delta I_j^k + O\left(\frac{1}{2}\frac{\partial^2 T}{\partial I_{j_1}\partial I_{j_2}}\delta I_{j_1}\delta I_{j_2}\right) z\,.$$

Keeping the first term:

$$\Rightarrow T\left(\bar{I}_k\right) \cong \widehat{\theta}_{I^*} + \left(\delta\widehat{\theta}_{i_1} + \delta\widehat{\theta}_{i_2} + \delta\widehat{\theta}_{i_3} \cdots + \delta\widehat{\theta}_{i_N}\right)\,, \tag{3.29}$$

where

$$\delta\widehat{\theta}_j = \left.\frac{\partial T}{\partial I_j}\right|_{I=I^*} \delta I_j^k\,. \tag{3.30}$$

That is, $\delta\widehat{\theta}_j$ is the change in the landmark location estimate due to a change in the quantized value for pixel j. Using Equations 3.25 and 3.26, $\widetilde{\theta}_k$ can be written as:

$$\begin{aligned}\widetilde{\theta}_k = \widetilde{\theta}_{\{i_1,i_2,i_3,\ldots,i_N\}} &= T(I^*) - T(\bar{I}_k) = \theta_{I^*} - \widehat{\theta}_k \\ &= \left(\theta_{I^*} - \widehat{\theta}_{I^*} - \left.\frac{\partial T}{\partial I}\right|_{I=I^*} \delta I^k\right) \\ &= \left(\widetilde{\theta}_{I^*} - \left.\frac{\partial T}{\partial I}\right|_{I=I^*} \delta I^k\right) \\ &= \left[\widetilde{\theta}_{I^*} - \left(\delta\widehat{\theta}_{i_1} + \delta\widehat{\theta}_{i_2} + \delta\widehat{\theta}_{i_3} \cdots + \delta\widehat{\theta}_{i_N}\right)\right]\,.\end{aligned} \tag{3.31}$$

Working in error coordinates, Equation 3.31 can be written as:

$$\widetilde{\theta}_k = \widetilde{\theta}_{\{i_1,i_2,i_3,\ldots,i_N\}} = \left(\delta\widehat{\theta}_{i_1} + \delta\widehat{\theta}_{i_2} + \delta\widehat{\theta}_{i_3} \cdots + \delta\widehat{\theta}_{i_N}\right)\,. \tag{3.32}$$

Substituting Equation 3.32 into Equation 3.24, we get:

$$E\left\{\widetilde{\theta}\widetilde{\theta}^T\right\} \tag{3.33}$$

$$= \sum_{i_1=1}^{M^{(1)}}\sum_{i_2=1}^{M^{(2)}}\sum_{i_3=1}^{M^{(3)}}\cdots\sum_{i_N=1}^{M^{(N)}} P_{i_1}P_{i_2}P_{i_3}\cdots P_{i_N}\left(\delta\widehat{\theta}_{i_1}\delta\widehat{\theta}_{i_2}\delta\widehat{\theta}_{i_3}\cdots\delta\widehat{\theta}_{i_N}\right)\left(\delta\widehat{\theta}_{i_1}\delta\widehat{\theta}_{i_2}\delta\widehat{\theta}_{i_3}\cdots\delta\widehat{\theta}_{i_N}\right)^T.$$

Equation 3.33 is the linear covariance approximation desired. Note that this formulation has linear computational complexity. This equation can be implemented by recursive induction, as follows:

Let's define the covariance by considering the first pixel as:

$$C_1 = \sum_{i_1=1}^{M^{(1)}} P_{i_1}\delta\widehat{\theta}_{i_1}\delta\widehat{\theta}_{i_1}^T\,, \tag{3.34}$$

which means that the covariance C_2 due to the contributions of pixels 1 and 2 is:

$$\begin{aligned} C_2 &= \sum_{i_1=1}^{M^{(1)}} \sum_{i_2=1}^{M^{(2)}} P_{i_1} P_{i_2} \left(\delta\widehat{\theta}_{i_1} + \delta\widehat{\theta}_{i_2}\right) \left(\delta\widehat{\theta}_{i_1} + \delta\widehat{\theta}_{i_2}\right)^T \\ &= \sum_{i_1=1}^{M^{(1)}} P_{i_1} \sum_{i_2=1}^{M^{(2)}} P_{i_2} \left(\delta\widehat{\theta}_{i_1}\delta\widehat{\theta}_{i_1}^T + \delta\widehat{\theta}_{i_1}\delta\widehat{\theta}_{i_2}^T + \delta\widehat{\theta}_{i_2}\delta\widehat{\theta}_{i_1}^T + \delta\widehat{\theta}_{i_2}\delta\widehat{\theta}_{i_2}^T\right) \\ &= \sum_{i_1=1}^{M^{(1)}} P_{i_1}\delta\widehat{\theta}_{i_1}\delta\widehat{\theta}_{i_1}^T + \sum_{i_1=1}^{M^{(1)}} P_{i_1} \sum_{i_2=1}^{M^{(2)}} P_{i_2} \left(\delta\widehat{\theta}_{i_1}\delta\widehat{\theta}_{i_2}^T + \delta\widehat{\theta}_{i_2}\delta\widehat{\theta}_{i_1}^T\right) + \sum_{i_1=1}^{M^{(1)}} P_{i_1} \sum_{i_2=1}^{M^{(2)}} P_{i_2}\delta\widehat{\theta}_{i_2}\delta\widehat{\theta}_{i_2}^T \\ &= \sum_{i_1=1}^{M^{(1)}} P_{i_1}\delta\widehat{\theta}_{i_1}\delta\widehat{\theta}_{i_1}^T + \sum_{i_1=1}^{M^{(1)}} P_{i_1}\delta\widehat{\theta}_{i_1} \sum_{i_2=1}^{M^{(2)}} P_{i_2}\delta\widehat{\theta}_{i_2}^T + \sum_{i_2=1}^{M^{(2)}} P_{i_2}\delta\widehat{\theta}_{i_2} \sum_{i_1=1}^{M^{(1)}} P_{i_1}\delta\widehat{\theta}_{i_1}^T + \sum_{i_2=1}^{M^{(2)}} P_{i_2}\delta\widehat{\theta}_{i_2}\delta\widehat{\theta}_{i_2}^T , \end{aligned} \tag{3.35}$$

which in turn can be expressed as

$$C_2 = C_1 + E\left\{\widehat{\theta}_{i_1}\right\} E\left\{\widehat{\theta}_{i_2}^T\right\} + E\left\{\widehat{\theta}_{i_2}\right\} E\left\{\widehat{\theta}_{i_1}^T\right\} + \sum_{i_2=1}^{M^{(2)}} P_{i_2}\delta\widehat{\theta}_{i_2}\delta\widehat{\theta}_{i_2}^T . \tag{3.36}$$

Finally, by induction, the covariance C_N due to the contributions of pixels $1, \ldots, N$ can be calculated from the covariance C_{N-1}:

$$\begin{aligned} C_N &= E\left\{\operatorname{Cov}\widetilde{\theta}\right\} \\ &= C_{N-1} + E\left\{\delta\widehat{\theta}_{i_1} + \delta\widehat{\theta}_{i_2} + \cdots + \delta\widehat{\theta}_{i_{N-1}}\right\} E\left\{\delta\widehat{\theta}_{i_N}^T\right\} \\ &\quad + E\left\{\delta\widehat{\theta}_{i_N}\right\} E\left\{\left(\delta\widehat{\theta}_{i_1} + \delta\widehat{\theta}_{i_2} + \cdots + \delta\widehat{\theta}_{i_{N-1}}\right)^T\right\} \\ &\quad + \sum_{i_N=1}^{M^{(N)}} P_{i_N}\delta\widehat{\theta}_{i_N}\delta\widehat{\theta}_{i_N}^T , \end{aligned} \tag{3.37}$$

where C_N is the covariance of the landmark location estimate ${}^{p}\widehat{P}_o$, considering the uncertainty introduced by pixels $1, \ldots, N$, and considering intensity quantization.

3.2.2 Landmark Location Estimator Confidence Interval

The procedure used to compute the landmark location estimator confidence interval is slightly different to the one used to compute the CRLB confidence interval. Practical estimators exhibit bias; that is, the expected value of the estimate is not equal to the real location of the image of the center of the landmark. Furthermore, in general these biases are not constant. This bias must be considered to determine the landmark location estimator confidence interval.

Figure 3.2 illustrates the bias and covariance computed at seven points for a particular landmark and estimator. The figure is made in this way, as each of seven points ${}^{p}P_o^*$, shown as crosses in the figure, the landmark mark is simulated,

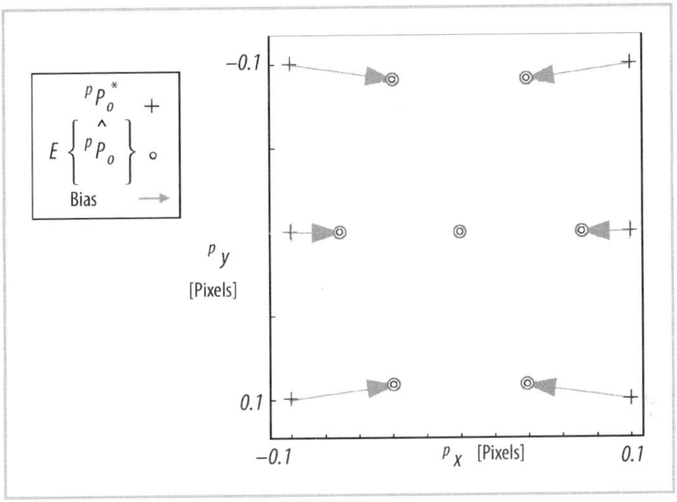

Figure 3.2. Estimator uncertainty contours at different positions estimates

the illuminance function is generated and J^* is computed. This is passed to the estimator, which gives:

$$ {}^p\widehat{P}_o^* = T_c\left(I^*\right) , \tag{3.38}$$

where $T_c()$ is the binary image centroid estimator. The vectors in Figure 3.2 show the bias. Notice that the bias is zero when ${}^pP_o^*$ falls at the center of the pixel. Finally an ellipse is drawn according to

$$\left({}^p\widehat{P}_o - {}^p\widehat{P}_o^*\right)^T \Sigma^{-1}\left({}^p\widehat{P}_o - {}^p\widehat{P}_o^*\right) = 5.99 , \tag{3.39}$$

where $\Sigma = C_N$ is the covariance matrix computed, and $\chi^2 = 5.99$ corresponds to the 95% confidence interval when two parameters are estimated.

The estimator uncertainty cumulative probability function at a distance R from ${}^pP_o^*$ is defined as:

$$\begin{aligned} F^T_{{}^p\widehat{P}_o}\left\{R; E\left\{{}^pP_o^*\right\} - {}^p\widehat{P}_o\right\} &= p\left(\left\|{}^p\widehat{P}_o - {}^pP_o^*\right\| < R\right) \\ &= \int\limits_{r=0}^{R}\int\limits_{\alpha=0}^{2\pi} rG\left({}^pP; E\left\{{}^p\widehat{P}_o\right\} - {}^pP_o^*, \Sigma_T\right) d\alpha\, dr , \end{aligned} \tag{3.40}$$

where $G\left({}^pP; {}^pP_o^*, \Sigma_{\text{CRLB}}\right)$ is a Gaussian probability function centered on $E\left\{{}^p\widehat{P}_o\right\} - {}^pP_o^*$ and covariance Σ_T:

$$G\left({}^pP\right) = \frac{1}{2\pi\sqrt{|\Sigma_T|}} e^{-\frac{1}{2}\left({}^pP - E\left\{{}^p\widehat{P}_o\right\} + {}^pP_o^*\right)^T \Sigma_T^{-1}\left({}^pP - E\left\{{}^p\widehat{P}_o\right\} + {}^pP_o^*\right)} , \tag{3.41}$$

where $|\Sigma_T|$ is the determinant of covariance matrix Σ_T. Consequently, the estimator uncertainty radial probability density function for a given position estimate ${}^pP_o^*$ is defined as (Peebles 2001):

$$f_{{}^p\widehat{P}_o}\left(R; E\left\{{}^pP^*_o\right\} - {}^p\widehat{P}_o\right) = \frac{dF_{{}^p\widehat{P}_o}\left\{R; E\left\{{}^pP^*_o\right\} - {}^p\widehat{P}_o\right\}}{dR} , \tag{3.42}$$

and correspondingly the estimator confidence interval for this estimate is:

$$C_n\left({}^pP^*_o\right) = \min\left\{ R \middle| \int_{r=0}^{R} f_{{}^p\widehat{P}_o}\left(r; E\left\{{}^pP^*_o\right\} - {}^p\widehat{P}_o\right) dr = n \right\} , \quad 0 < n \leq 1 , \tag{3.43}$$

where n reflects the confidence level (*e.g.*, $n = 95\%$ or $n = 99\%$), and C_n is the radius of a circle within which ${}^p\widehat{P}$ will lie in the fraction n of cases.

Finally, the overall confidence interval is created by the integration of Equation 3.43 in the $\pm 1/2$ pixel region Ω:

$$\overline{C_n} = \iint_{\Omega} C_n\left({}^pP^*_o\right) dx\, dy . \tag{3.44}$$

Monte Carlo integration is used to evaluate Equation 3.44, to overcome the challenge of a rapidly varying integrand. In Chapter 8, results are presented where the 3-D integrations of Equations 3.43 and 3.44 are evaluated numerically to determine the Cramér–Rao lower bound on uncertainty for a range of configurations.

3.3 Discussion

The methods developed here employ a mix of analytic results where tractable and numerical methods where necessary. Using these tools, the Cramér–Rao lower bound can be computed for a given landmark and imager configuration, independent of the estimator used , and the uncertainty of practical estimators can be determined. This chapter, and indeed this entire work, has focused on circular landmarks, but with suitable extension of the methods, the analysis can be applied to arbitrary landmark shapes. The calculation of $\left(\partial J/\partial\, {}^pP^*_o\right)$, for example, is achieved by integrating the illuminance along the edges of the sensitive areas, which can be done whenever the illuminance function can be formed.

In the next chapter, two model-based landmark location estimators are developed. Following that, in Chapters 5 and 6, the mechanics for implementing the results of this chapter are laid out. This is then followed, in Chapter 8, by the analysis of a circular landmark and practical estimators.

4 Model-based Landmark Location Estimators

The most important feature and information-rich portion of a landmark is its edge. The landmark edge is smoothed by both the finite sensitive area of the pixel and the illuminance function. The illuminance function of a circular landmark, rendered in a 3-D plot such as Plate 4 (see Color Section, page 147), has the look of a tepuy[1] or table mountain.

Two model-based landmark location estimators are presented here. These estimators reduce bias by explicitly modeling the illuminance function given by the landmark, and they reduce uncertainty by exploiting the intermediate grayscale values given by smoothing at the edge of the landmark. These algorithms are inspired, in part, by the Nalwa–Binford algorithm for precise edge location (Nalwa and Binford 1986). This algorithm locates a line-edge by modeling the smooth transition between light on one side and dark on the other.

The ellipsoidal contour algorithm, described in Section 4.1, operates by fitting an ellipse to points produced by the contour routine in MATLAB®. Interior and exterior intensities are first detected, so that the contour level can be selected as the mean intensity. This corresponds approximately to the midway point of the transition between the interior and exterior of the landmark. The ellipsoidal contour model has five parameters, corresponding to an ellipse center, ${}^{p}\widehat{P}_0$, radius, eccentricity, and orientation. The algorithm was first developed to seed the parameter values for the more complex Butterworth tepuy algorithm; but to our surprise it gave an analytic and experimental performance indistinguishable from the more complex algorithm. An explanation for this is considered after the estimators are presented.

The Butterworth tepuy algorithm, described in Section 4.2, was developed to capture a parametric model of the illuminance of a circular landmark that can be fit to the measured digital image. The fit can be carried out with, for example, the Newton–Raphson method. The model is parametric in the sense that it is expressed in closed form. While it captures the width of the transition region as a parameter, just like the Nalwa–Binford algorithm, the model does not rely on numerical values for pixel sensitive area or smoothing radius.

As will be shown through both analysis and experiment, the two model-based estimators significantly reduce landmark location uncertainty. Indeed, they oper-

[1] Tepuy means "table mountain" in the native language of the Pemon tribes located in the Gran Sabana, Venezuela.

ate near the Cramér–Rao lower bound. These estimators model and locate circular landmarks. However, other landmark shapes can be accommodated by developing a suitable model for the expected landmark image.

4.1 Ellipsoidal Contour Landmark Location Estimator

The edge of any feature in an image is inherently smoothed by the finite sensitive area of each pixel and by the smoothing effect of diffraction and defocus. In this way, pixel values near the edge take intermediate values between the intensity of the interior of the landmark and that of the background, and the exact value taken in each pixel depends on the landmark location.

For this discussion, the landmark interior is taken to give an image with a discrete analog intensity of $J^* = A_I$, and the background is taken to give $J^* = A_B$. To apply the ellipsoidal contour algorithm, the landmark must have first been detected in the image and approximate values for A_I and A_B obtained. Detecting the landmark and obtaining interior and background intensity values are, of course, challenging problems in their own right, but are not considered here. Experience suggests that the algorithm is reliable so long as the uncertainty in A_I and A_B is small compared to the difference between them.

Starting with an approximate knowledge of the location of the landmark and the intensity values A_I and A_B, the algorithm proceeds by selecting pixels corresponding to the landmark and its immediate exterior; determining the intensity level for the contour, $A_c = \left(A_I + A_B\right)/2$, and determining points along the A_c contour by interpolation. The MATLAB® contour routine was used in this study. This routine uses a linear interpolation method that works in the following way: at each pixel, the pixel intensity and those of the four immediate neighbors are evaluated, and edges are identified with values bracketing A_c; on each identified edge, linear interpolation is used to determine the point in the image lying on the A_c contour. An example is seen in Figure 4.1, where a contour is illustrated on the digital image. For this example, the contour interpolation algorithm detected $N_c = 28$ edges bracketed by $\overline{I}$ values above and below A_c, giving 28 points ${}^{P}P_a = \left[x_j, y_j\right]^T$, with $j = 1, \ldots, N_c$. In this example, although the diameter of the landmark is approximately six pixels, 28 contour points are determined by interpolation, providing 56 data values that are used by the ellipsoid contour algorithm. There is some redundancy in the data, since there are only 40 pixels that bracket the contour. Interestingly, for this landmark of diameter six pixels, there are more pixels that contribute to the determination of the contour ($N = 40$) than there are pixels on the interior of the landmark ($N \approx 3^2\pi \approx 28$).

A five-parameter ellipsoid model is fit to the contour points according to:

$$\widetilde{z}_j \begin{bmatrix} a & b \\ b & c \end{bmatrix} \widetilde{z}_j^T = 1 \qquad \widetilde{z}_j = \begin{bmatrix} x_j - x_o \\ y_j - y_o \end{bmatrix} \tag{4.1}$$

where a, b, c are the parameters of the ellipse centered at coordinates x_o, y_o, and x_j, y_j is the location of the jth contour point, $j = 1, \ldots, N_c$.

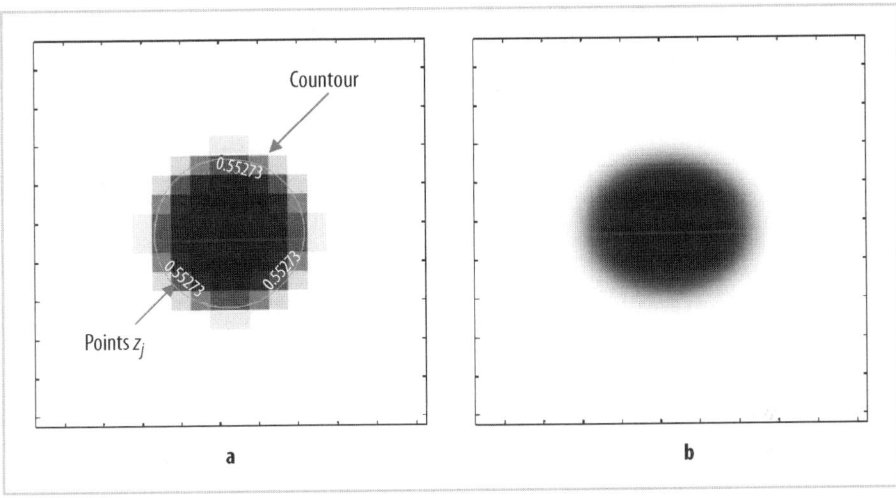

Figure 4.1. Example of ellipsoidal contour around a landmark. Subplot **b** shows the illuminance function of the image of the landmark. Subplot **a** shows the corresponding digital image and contour fit to the data.

A nonlinear least squares estimation for parameters $\theta = [a, b, c, x_o, y_o]^T$ is done for this model, using the Newton–Raphson method. As a nonlinear estimation technique, the Newton–Raphson algorithm requires initial parameter values, which should be as close as is practical to the true values. In principle the algorithm can fail if the initial, or "seed," values are too far from the true solution, but in this case suitable initial parameters are easily determined:

- Seed values for x_o, y_o are determined by applying the centroid algorithm to the landmark;
- A linear fit is used to find seed values for the ellipsoid parameters a, b, c by evaluating the $3 \times N_c$ system of equations:

$$1 = [a \;\; b \;\; c] \begin{bmatrix} (x_j - x_0)^2 \\ (x_j - x_0)(y_j - y_0) \\ (y_j - y_0)^2 \end{bmatrix} . \tag{4.2}$$

Once the seed values are determined, the Newton–Raphson algorithm is run to minimize the error:

$$\varepsilon = \sum_{j=1}^{N_c} \left(\tilde{z}_j \begin{bmatrix} a & b \\ b & c \end{bmatrix} \tilde{z}_j^T - 1 \right)^2 ; \qquad \tilde{z}_j = \begin{bmatrix} x_j - x_o \\ y_j - y_o \end{bmatrix} . \tag{4.3}$$

The estimate for the center of the image of the landmark is taken directly from the fitted parameters x_o, y_o. Figure 4.1 shows a contour established on the digital image of a circular landmark (a) from the illuminance function displayed in (b). The points $z_j = \begin{bmatrix} x_j - x_o \\ y_j - y_o \end{bmatrix}$ correspond to pixels that belong to the contour established.

4.2 Butterworth Tepuy Landmark Location Estimator

This model-based estimator is inspired by the shape of a tepuy or table mountain, as it models the 3-D shape of the illuminance function that corresponds to a circular landmark. As described in Chapter 2, the diffraction phenomenon is modeled as the convolution of the optical flux function $H(x, y)$ and a 2-D Gaussian kernel to produce the illuminance function $E(x, y)$. The result of convolving the Gaussian kernel with a circular landmark will be called a Gaussian tepuy (see Plate 5, part a, in the Color Section, page 148).

The closed-form equation of the Gaussian tepuy does not exist; for this reason it is necessary to approximate its mathematical formulation with another function (Proakis and Manolakis 1996). A suitable approximation is the expression of the frequency magnitude of a 2-D Butterworth filter, scaled appropriately. The expression of the Butterworth tepuy is:

$$\begin{aligned} E(x,y) &\approx E_{\mathrm{BT}}\left(x,y;a,b,c,x_o,y_o,L_{\max},L_{\min},s_b\right) \\ &= \frac{L_{\max}-L_{\min}}{1+\left(\frac{\left(x-x_o\right)^2+\left(y-y_o\right)^2}{R(x,y)^2}\right)^{\frac{R(x,y)}{s_b}}}+L_{\min}\,, \end{aligned} \tag{4.4}$$

where $E(x, y)$ is the true illuminance function, $E_{\mathrm{BT}}(\cdot)$ is the Butterworth tepuy model, and $R(x, y)$ is the radius of an ellipse along the vector $\boldsymbol{V} = [(x-x_o), (y-y_o)]$, defined as:

$$R(x,y) = R(x,y;a,b,c,x_o,y_o) = \sqrt{\frac{\frac{\left(x-x_o\right)^2}{\left(y-y_o\right)^2}+1}{a\frac{\left(x-x_o\right)^2}{\left(y-y_o\right)^2}+2b\frac{\left(x-x_o\right)}{\left(y-y_o\right)}+c}}\,. \tag{4.5}$$

Plate 5 shows a comparison between a circular landmark smoothed by a Gaussian diffraction kernel with variance σ_b^2 and a Butterworth tepuy.

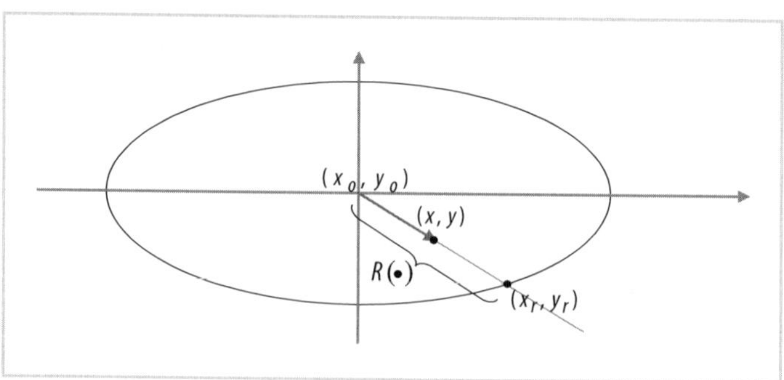

Figure 4.2. Ellipsoidal radius

The Butterworth tepuy landmark location estimator fits a surface to the image of a circular landmark using the model expressed in Equation 4.4. The parameters a, b, c are the parameters of the ellipse centered at coordinates x_o, y_o. $L_{\max}$ and $L_{\min}$ are the highest and lowest intensity levels of the tepuy and s_b is related to the transition between $L_{\max}$ and $L_{\min}$, also called the skirt width.

In this model, a nonlinear least squares estimator with parameters $\widehat{\theta} = \left[a, b, c, x_o, y_o, L_{\max}, L_{\min}, s_b\right]^T$ is solved using the Newton–Raphson method. The seed for the nonlinear fit is established as follows:

- The ellipsoidal contour estimator is used to find the seed for a, b, c, x_o, y_o.
- The intensity of the pixel nearest position x_o, y_o is used as the seed for $L_{\max}$.
- The lowest intensity level in the neighborhood of the landmark is used as the seed for $L_{\min}$ (or the highest intensity level if the landmark is dark on a light background).
- The width of one pixel is used as the seed for s_b.

The expression for the ellipsoidal radius requires special attention. It must be written in such a way that the derivatives of the data with respect to the Butterworth tepuy parameters can be formed for the Newton–Raphson calculation. The expression for the ellipsoidal radius is justified in the following manner. From Figure 4.2, $R\left(x, y\right)$ can be written as:

$$R\left(x, y\right) = \sqrt{\left(x_r(x, y) - x_o\right)^2 + \left(y_r(x, y) - y_o\right)^2} \,. \tag{4.6}$$

The slope of the line segment formed between $\left(x_o, y_o\right)$ and $\left(x_r, y_r\right)$ is:

$$\frac{y_r(x, y) - y_o}{x_r(x, y) - x_o} = \frac{y - y_o}{x - x_o} = \frac{1}{S} \tag{4.7}$$

$$\Rightarrow \left(x_r(x, y) - x_o\right) = S\left(y_r(x, y) - y_o\right) \quad \text{or} \quad S(x, y) = \frac{\left(x_r(x, y) - x_o\right)}{\left(y_r(x, y) - y_o\right)} \,. \tag{4.8}$$

From the definition of the ellipse:

$$a\left(x_r(x, y) - x_o\right)^2 + 2b\left(x_r(x, y) - x_o\right)\left(y_r(x, y) - y_o\right) + c\left(y_r(x, y) - y_o\right)^2 = 1 \,. \tag{4.9}$$

Defining $\widetilde{P}$ as:

$$\widetilde{P} = \begin{bmatrix} x_r(x, y) \\ y_r(x, y) \end{bmatrix} - \begin{bmatrix} x_o \\ y_o \end{bmatrix} , \tag{4.10}$$

Equation 4.9 can be written as:

$$\widetilde{P}^T \begin{bmatrix} a & b \\ b & c \end{bmatrix} \widetilde{P} = 1 \,, \tag{4.11}$$

or equivalently

$$\widetilde{P}^T Q \widetilde{P} = \left\| \widetilde{P} \right\|_Q = 1 \,, \tag{4.12}$$

where

$$Q = \begin{bmatrix} a & b \\ b & c \end{bmatrix} . \tag{4.13}$$

Substituting Equation 4.8 into Equation 4.9, we get

$$\begin{aligned} & aS^2 \left(y_r(x,y) - y_o \right)^2 + 2bS \left(y_r(x,y) - y_o \right)^2 + c \left(y_r(x,y) - y_o \right)^2 = 1 \\ & \Rightarrow \left(y_r(x,y) - y_o \right)^2 = \frac{1}{aS^2 + 2bS + c} \,. \end{aligned} \tag{4.14}$$

From Equations 4.8 and 4.14:

$$\Rightarrow \left(x_r(x,y) - x_o \right)^2 = \frac{S^2}{aS^2 + 2bS + c} \,. \tag{4.15}$$

Substituting Equations 4.14 and 4.15 into Equation 4.6, we get:

$$R(x,y) = \sqrt{\frac{S(x,y)^2}{aS(x,y)^2 + 2bS(x,y) + c} + \frac{1}{aS(x,y)^2 + 2bS(x,y) + c}} \,. \tag{4.16}$$

Finally, substituting Equation 4.8 into Equation 4.16, we obtain:

$$R(x,y) = R(x,y;a,b,c,x_o,y_o) = \sqrt{\frac{\dfrac{(x-x_o)^2}{(y-y_o)^2} + 1}{a\dfrac{(x-x_o)^2}{(y-y_o)^2} + 2b\dfrac{(x-x_o)}{(y-y_o)} + c}} \,. \tag{4.17}$$

Equation 4.17 can be expressed in a more compact form:

$$\begin{aligned} R(x,y) = R(x,y;a,b,c,x_o,y_o) &= \sqrt{\frac{(x-x_o)^2 + (y-y_o)^2}{a(x-x_o)^2 + 2b(x-x_o)(y-y_o) + c(y-y_o)^2}} \\ &= \sqrt{\|\widetilde{P}\|^2 / \|\widetilde{P}\|_Q^2} \end{aligned} \tag{4.18}$$

or

$$R(x,y) = R(x,y;a,b,c,x_o,y_o) = \|\widetilde{P}\| / \|\widetilde{P}\|_Q \,. \tag{4.19}$$

Using Equation 4.19, the expression for the Butterworth tepuy, Equation 4.4, can be further simplified by substitution:

$$E_{\mathrm{BT}}(x, y; a, b, c, x_o, y_o, L_{\max}, L_{\min}, s_b) = \frac{L_{\max} - L_{\min}}{1 + \left(\|\widetilde{P}\|_Q^2\right)^{\frac{\|\widetilde{P}\| / \|\widetilde{P}\|_Q}{s_b}}} + L_{\min} \,. \tag{4.20}$$

Using Equation 4.20, the parameters of the Butterworth tepuy model are fit by using the Newton–Raphson method to minimize

$$\varepsilon = \sum_{j=1}^{N_{\mathrm{p}}} \left(\bar{I}(j) - E_{\mathrm{BT}}(\widehat{\theta})\right)^2 \,, \tag{4.21}$$

where $\widehat{\theta} = \left[a, b, c, x_o, y_o, L_{\max}, L_{\min}, s_b\right]^T$ is the parameter vector, j is the pixel index, and N_{p} is the number of pixels used for the estimation. The estimate for the center of the image of the landmark is taken directly from the fitted parameters, ${}^{P}\widehat{P}_{\underset{L}{o}} = \begin{bmatrix} x_o \\ y_o \end{bmatrix}$.

4.3 Discussion

Two model-based estimators have been presented. They operate very differently. The ellipsoidal contour algorithm fits the parameters of an ellipse to contour points obtained by interpolating the pixel data. The Butterworth tepuy algorithm fits the intensity data directly. In spite of these differences and a significant difference in computational cost (with the ellipsoidal contour being the faster to compute), the two model-based estimators show closely comparable performance. In what follows, these location estimators will be evaluated with the analysis methods derived in Chapter 3 using the tools described in Chapter 6. The location estimators will also be tested with the experiment described in Chapter 7.

5 Two-dimensional Noncollocated Numerical Integration

In order to apply either the Cramér–Rao bound of Section 3.1 or the analytic tools of Section 3.2, the integral of Equation 2.27 must be performed many times. Additionally, the derivatives of Equation 3.12 are obtained by performing a 1-D integration along the edges of the sensitive areas. In each case, the integration is of the illuminance function $E({}^ix, {}^iy)$. Because the illuminance function is given as a convolution with a Gaussian kernel, no closed-form expression is available for $E({}^ix, {}^iy)$, which is determined numerically on a high-resolution sample grid by numerical convolution of the optical flux function with the smoothing operator. The integrations of Equations 2.27 and 3.12 are then performed numerically on the high-resolution grid.

The conditions for the numerical integration are illustrated qualitatively in Figure 5.1. The points in Figure 5.1 represent the sample points in a high-resolution grid; these are sometimes called the "knots" in the literature on multidimensional integration (*e.g.*, Krommer and Ueberhuber 1998; Stroud 1971). These points with spacing h_x and h_y have periodic spacing that must have a resolution sufficient to represent the smoothing operator; that is $h_x, h_y \ll \sigma_b$. The region of integration is the sensitive area of the pixel, illustrated in dark gray in Figure 5.1. Suitable values for h_x and h_y are found by the error analysis presented in Appendix C.

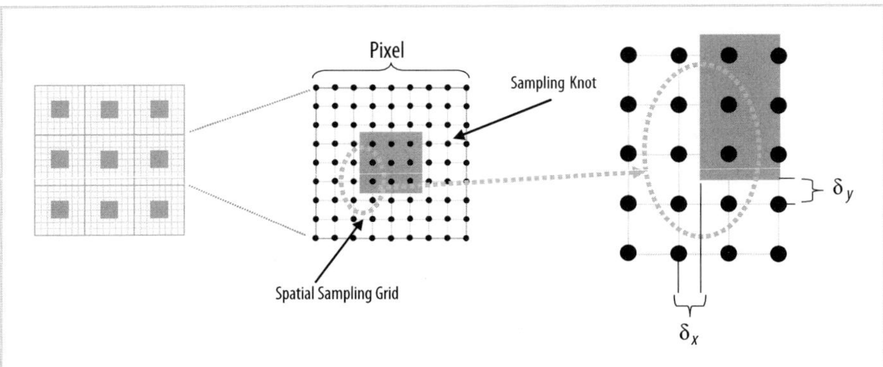

Figure 5.1. Integration region for the spatial sampling. The sensitive area (*dark gray*) is the region of integration, and it is not collocated with the sampling grid.

Many multivariable numerical integration formulae (called cubature formulae) are available for regions in $\mathfrak{R}^2$ [Krommer and Ueberhuber 1998], but they all require that the region of integration be collocated with the knots. In Figure 5.1, the region of integration (dark gray) is not collocated with the high-resolution sampling grid. Of course, one may choose a sampling grid that is collocated with the sensitive area. However, for the purposes of this study, two problems arise. First, the grid extends over many pixels, and a grid aligned with one sensitive area might not be aligned with another. Second, and more fundamentally, one would like to analyze the landmark at many locations on the imager. Analysis could be performed at each value of ${}^{p}P_{o}^{*}$ by reforming the high-resolution grid and recomputing the convolution to obtain the illuminance function. However, it is more efficient to form the illuminance function once, and model many landmark locations by shifting the sensitive areas with respect to the illuminance function. In this way, the 2-D convolution needed to generate the illuminance function is performed only once. One requirement to achieve this efficiency, however, is that the integration of the illuminance function can handle the case in which the sensitive area is not collocated with the high-resolution grid.

Two-dimensional numerical integration where the region of integration is not collocated with the knots is the focus of this chapter. A suitable cubature formula is developed here, including details of its computation. Proof that the approximation error is well behaved is provided in Appendix C. For the reader who is willing to take it on faith that a proper 2-D integration rule has been found, perusal of this chapter can be deferred until the implementation details become of interest.

5.1 Noncollocated Simpson Rule: 1-D

In order to numerically integrate the illuminance function across each sensitive area without requiring that the sensitive area be collocated with the high-resolution grid, a new noncollocated 2-D numerical integration method has been developed. The integration formula is based on a multivariable version of the Simpson rule, which has been modified to allow the integration limits to lie anywhere inside the high-resolution grid. The use of the Simpson formula framework brings several advantages:

- A product formula over a rectangular support is straightforward to implement in comparison with other cubature formulae (Krommer and Ueberhuber 1998).
- In the model used, the integrand is known only in a uniform lattice of points or knots of the sampling grid (see Figure 5.1). Due to the convolution process, the illuminance function is only known at the knots of the sampling grid (the lack of complete knowledge of the integrand within the integration region blocks the use of the Gauss product formulae; Stroud 1971).
- The integration error is of $O(h^3)$.
- All of the formula coefficients are positive, which implies good numerical stability (Stroud 1971).

The Simpson rule selected was the popular second-order 1/3 rule (Stroud 1971).

The derivation of the original 1-D Simpson rule is made straightforward through the use of the Lagrange interpolation polynomials. The Lagrange form of the quadratic interpolation polynomial for a continuous function $g(z)$ is:

$$P_2(z) = L_o(z)g(z_o) + L_1(z)g(z_1) + L_2(z)g(z_2)\,, \tag{5.1}$$

where

$$\{z_o, z_1, z_2 \in \Re;\ z_o < z_1 < z_2\}\,, \tag{5.2}$$

and

$$\begin{aligned} L_o(z) &= \frac{(z-z_1)(z-z_2)}{(z_o-z_1)(z_o-z_2)} \\ L_1(z) &= \frac{(z-z_0)(z-z_2)}{(z_1-z_0)(z_1-z_2)} \\ L_2(z) &= \frac{(z-z_0)(z-z_1)}{(z_2-z_0)(z_2-z_1)} \end{aligned} \tag{5.3}$$

are the Lagrange basis functions (Stewart 1994). It is easy to see by direct substitution that z_0, z_1, and z_2 are points such that $P_2(z_i) = g(z_i)$. Using the above quadratic interpolation, the integral of the function $g(z)$ in the interval $[z_o, z_2]$ can be approximated as:

$$\int_{z_o}^{z_2} g(z)\,dz \approx \int_{z_o}^{z_2} P_2(z)\,dz$$

$$\begin{aligned} &= g(z_o)\int_{z_o}^{z_2} L_0(z)\,dz + g(z_1)\int_{z_o}^{z_2} L_1(z)dz + g(z_2)\int_{z_o}^{z_2} L_2(z)\,dz \\ &= \frac{g(z_o)}{(z_o-z_1)(z_o-z_2)}\left[\frac{z^3}{3} - \frac{(z_1+z_2)}{2}z^2 + z_1z_2z\right]_{z_o}^{z_2} \\ &+ \frac{g(z_1)}{(z_1-z_o)(z_1-z_2)}\left[\frac{z^3}{3} - \frac{(z_o+z_2)}{2}z^2 + z_oz_2z\right]_{z_o}^{z_2} \\ &+ \frac{g(z_2)}{(z_2-z_o)(z_2-z_1)}\left[\frac{z^3}{3} - \frac{(z_o+z_1)}{2}z^2 + z_oz_1z\right]_{z_o}^{z_2} \\ &= \frac{g(z_o)}{(z_o-z_1)(z_o-z_2)}\left[\frac{\left(z_o^3-z_2^3\right)}{3} + \frac{(z_1+z_2)\left(z_o^2-z_2^2\right)}{2} - z_1z_2\left(z_o-z_2\right)\right] \\ &+ \frac{g(z_1)}{(z_1-z_o)(z_1-z_2)}\left[\frac{\left(z_o^3-z_2^3\right)}{3} + \frac{(z_o+z_2)\left(z_o^2-z_2^2\right)}{2} - z_oz_2\left(z_o-z_2\right)\right] \\ &+ \frac{g(z_2)}{(z_2-z_o)(z_2-z_1)}\left[\frac{\left(z_o^3-z_2^3\right)}{3} + \frac{(z_o+z_1)\left(z_o^2-z_2^2\right)}{2} - z_oz_1\left(z_o-z_2\right)\right] \end{aligned}$$

$$= \frac{g(z_o)}{(z_o - z_1)} \left[-\frac{(z_o^2 + z_o z_2 + z_2^2)}{3} + \frac{(z_1 + z_2)(z_o + z_2)}{2} - z_1 z_2 \right]$$
$$+ \frac{g(z_1)(z_o - z_2)}{(z_1 - z_o)(z_1 - z_2)} \left[-\frac{(z_o^2 + z_o z_2 + z_2^2)}{3} + \frac{(z_o + z_2)^2}{2} - z_o z_2 \right]$$
$$- \frac{g(z_2)}{(z_2 - z_1)} \left[-\frac{(z_o^2 + z_o z_2 + z_2^2)}{3} + \frac{(z_o + z_1)(z_o + z_2)}{2} - z_o z_1 \right] . \quad (5.4)$$

For the case where the points z_0, z_1, z_2 are equidistant:

$$z_2 - z_1 = z_1 - z_0 = h \,, \quad (5.5)$$

then Equation 5.4, after algebraic manipulation, reduces to

$$\int_{z_o}^{z_2} g(z)dz \approx \frac{h}{3} \left[g(z_o) + 4g(z_1) + g(z_2) \right] , \quad (5.6)$$

which is the well-known 1/3 Simpson rule.

For the case where the lower integration limit is not z_o but $z_o + \delta_z^L$ with $\delta_z^L < z_2 - z_o$, even when the function is known only at the points z_0, z_1, and z_2, it is possible to derive a modified Simpson rule. Using the Lagrange basis functions from Equation 5.3, we can write:

$$\int_{z_o+\delta_z^L}^{z_2} g(z)\,dz \approx \int_{z_o+\delta_z^L}^{z_2} P_2(z)\,dz$$
$$= g(z_o) \int_{z_o+\delta_z^L}^{z_2} L_0(z)dz + g(z_1) \int_{z_o+\delta_z^L}^{z_2} L_1(z)dz + g(z_2) \int_{z_o+\delta_z^L}^{z_2} L_2(z)\,dz$$
$$= \frac{g(z_o)}{(z_o - z_1)(z_o - z_2)} \left[\frac{z^3}{3} - \frac{(z_1 + z_2)}{2} z^2 + z_1 z_2 z \right]_{z_o+\delta_z^L}^{z_2}$$
$$+ \frac{g(z_1)}{(z_1 - z_o)(z_1 - z_2)} \left[\frac{z^3}{3} - \frac{(z_o + z_2)}{2} z^2 + z_o z_2 z \right]_{z_o+\delta_z^L}^{z_2}$$
$$+ \frac{g(z_2)}{(z_2 - z_o)(z_2 - z_1)} \left[\frac{z^3}{3} - \frac{(z_o + z_1)}{2} z^2 + z_o z_1 z \right]_{z_o+\delta_z^L}^{z_2} . \quad (5.7)$$

Using the condition presented in Equation 5.5, and after algebraic manipulation, Equation 5.7 becomes:

$$\int_{z_o+\delta_z^L}^{z_2} g(z)\,dz \approx \frac{h}{3}\left[g(z_o)\left(1 - 3\frac{\delta_z^L}{h} + \frac{9}{4}\left(\frac{\delta_z^L}{h}\right)^2 - \frac{1}{2}\left(\frac{\delta_z^L}{h}\right)^3\right) + g(z_1)\left(4 - 3\left(\frac{\delta_z^L}{h}\right)^2 + \left(\frac{\delta_z^L}{h}\right)^3\right) + g(z_2)\left(1 + \frac{3}{4}\left(\frac{\delta_z^L}{h}\right)^2 - \frac{1}{2}\left(\frac{\delta_z^L}{h}\right)^3\right)\right]. \tag{5.8}$$

Similarly, for the case where the upper integration limit is not z_2 but $z_2 - \delta_z^U$ with $\delta_z^U < z_2 - z_o$, even when the function is known only at the points z_o, z_1, and z_2, the modified Simpson rule will be:

$$\int_{z_o}^{z_2-\delta_z^U} g(z)\,dz \approx \frac{h}{3}\left[g(z_o)\left(1 + \frac{3}{4}\left(\frac{\delta_z^U}{h}\right)^2 - \frac{1}{2}\left(\frac{\delta_z^U}{h}\right)^3\right) + g(z_1)\left(4 - 3\left(\frac{\delta_z^U}{h}\right)^2 + \left(\frac{\delta_z^U}{h}\right)^3\right) + g(z_2)\left(1 - 3\frac{\delta_z^U}{h} + \frac{9}{4}\left(\frac{\delta_z^U}{h}\right)^2 - \frac{1}{2}\left(\frac{\delta_z^U}{h}\right)^3\right)\right]. \tag{5.9}$$

In order to derive the composite version of the modified Simpson rule, we proceed to divide the integration interval $[z_o, z_n]$ into an even number of subintervals n of size h (see Figure 5.2); that is:

$$[z_o, z_n] = [z_o, z_2], [z_2, z_4], \cdots [z_{n-2}, z_n;] \quad n \text{ even} \tag{5.10}$$

and

$$h = \frac{z_n - z_o}{n} = z_{i+1} - z_i, \quad i = (0, \ldots, n-1);\ n \text{ even}. \tag{5.11}$$

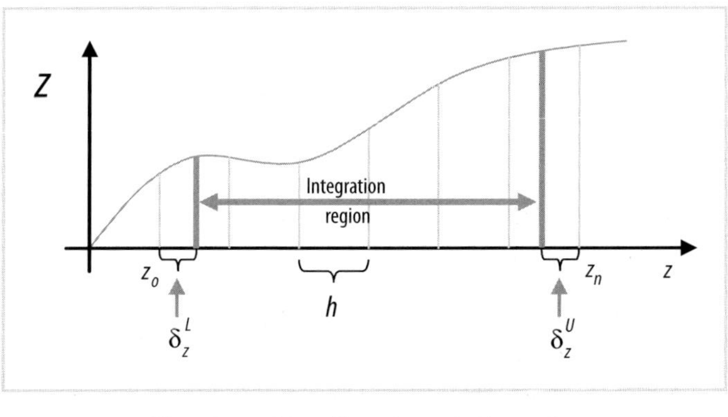

Figure 5.2. 1-D noncollocated numerical integration

With this we can write:

$$\int_{z_o+\delta_z^L}^{z_n-\delta_z^U} g(z)\,dz = \int_{z_o+\delta_z^L}^{z_2} g(z)\,dz + \int_{z_2}^{z_4} g(z)\,dz + \cdots + \int_{z_{n-1}}^{z_n-\delta_z^U} g(z)\,dz\,. \tag{5.12}$$

Substituting Equations 5.6, 5.8, and 5.9 into Equation 5.12, we obtain:

$$\begin{aligned}\int_{z_o+\delta_z^L}^{z_n-\delta_z^U} g(z)\,dz \approx \frac{h}{3}\Bigg[& g(z_o) + 4\sum_{i=0}^{n/2-1} g(z_{2i+1}) + 2\sum_{i=1}^{n/2-1} g(z_{2i}) + g(z_n) \\ & -\left\langle V\left(\frac{\delta_z^L}{h}\right), [g(z_o)\; g(z_1)\; g(z_2)]^T \right\rangle \\ & -\left\langle V\left(\frac{\delta_z^U}{h}\right), [g(z_n)\; g(z_{n-1})\; g(z_{n-2})]^T \right\rangle \Bigg]\,,\end{aligned} \tag{5.13}$$

where

$$V(\nu) = \left[\left(3\nu - \frac{9}{4}\nu^2 + \frac{1}{2}\nu^3\right)\;(3\nu^2 - \nu^3)\;\left(-\frac{3}{4}\nu^2 + \frac{1}{2}\nu^3\right)\right] \tag{5.14}$$

and $\langle \boldsymbol{a}, \boldsymbol{b} \rangle$ is the vector inner product operator.

Equation 5.13 is the 1-D noncollocated form of Simpson's integration rule. It is the well-known 1-D Simpson rule, but with the addition of two correction terms corresponding to the low and high integration limits.

5.2 Noncollocated Simpson Rule: 2-D

The derivation of the 2-D modified Simpson rule follows. The integral of the function $f(x,y)$ in the interval $\left\lfloor x_o + \delta_x^L, x_{n_x} - \delta_x^U \right\rfloor \times \left\lfloor y_o + \delta_y^L, y_{n_y} - \delta_y^U \right\rfloor$ can be written as:

$$\int_{y_o+\delta_y^L}^{y_{n_y}-\delta_y^U} \int_{x_o+\delta_x^L}^{x_{n_x}-\delta_x^U} f(x,y)\,dx\,dy = \int_{y_o+\delta_y^L}^{y_{n_y}-\delta_y^U} g(y)\,dy\,, \tag{5.15}$$

where

$$g(y) = \int_{x_o+\delta_x^L}^{x_{n_x}-\delta_x^U} f(x,y)\,dx\,. \tag{5.16}$$

Using Equation 5.13, Equation 5.16 can be approximated as:

$$\begin{aligned} g(y) \approx \frac{h_x}{3}\Bigg[& f(x_o,y) + 4\sum_{i=0}^{\frac{n_x}{2}-1} f(x_{2i+1},y) + 2\sum_{i=1}^{\frac{n_x}{2}-1} f(x_{2i},y) + f(x_{n_x},y) \\ & - \left\langle V\left(\frac{\delta_x^L}{h_x}\right), [f(x_o,y)\, f(x_1,y)\, f(x_2,y)]^T \right\rangle \\ & - \left\langle V\left(\frac{\delta_x^U}{h_x}\right), [f(x_{n_x},y)\, f(x_{n_x-1},y)\, f(x_{n_x-2},y)]^T \right\rangle \Bigg]\,, \end{aligned} \tag{5.17}$$

with

$$h_x = \frac{x_{n_x} - x_o}{n_x} = x_{i+1} - x_i, \quad i = (0,\ldots,n_x - 1);\; n_x \text{ even}\,. \tag{5.18}$$

Substituting Equation 5.17 into Equation 5.15, we obtain:

$$\begin{aligned} & \int_{y_o+\delta_y^L}^{y_{n_y}-\delta_y^U} \int_{x_o+\delta_x^L}^{x_{n_x}-\delta_x^U} f(x,y)\,dx\,dy \\ & \approx \int_{y_o+\delta_y^L}^{y_{n_y}-\delta_y^U} \frac{h_x}{3}\Bigg[f(x_o,y) + 4\sum_{i=0}^{\frac{n_x}{2}-1} f(x_{2i+1},y) + 2\sum_{i=1}^{\frac{n_x}{2}-1} f(x_{2i},y) + f(x_{n_x},y) \\ & - \left\langle V\left(\frac{\delta_x^L}{h_x}\right), [f(x_o,y)\, f(x_1,y)\, f(x_2,y)]^T \right\rangle \\ & - \left\langle V\left(\frac{\delta_x^U}{h_x}\right), [f(x_{n_x},y)\, f(x_{n_x-1},y)\, f(x_{n_x-2},y)]^T \right\rangle \Bigg] dy\,. \end{aligned} \tag{5.19}$$

Repeating the application of the Simpson rule to Equation 5.19:

$$\int_{y_o+\delta_y^L}^{y_{n_y}-\delta_y^U} \int_{x_o+\delta_x^L}^{x_{n_x}-\delta_x^U} f(x,y)\,dx\,dy \approx$$

$$\approx \frac{h_y}{3} \times$$

$$\times \left\{ \begin{array}{l}
\frac{h_x}{3} \left[\begin{array}{l} f(x_o,y_o) + 4 \sum_{i=0}^{\frac{n_x}{2}-1} f(x_{2i+1},y_o) + 2 \sum_{i=1}^{\frac{n_x}{2}-1} f(x_{2i},y_o) + f(x_{n_x},y_o) \\ - \left\langle V\left(\frac{\delta_x^L}{h_x}\right), \begin{bmatrix} f(x_o,y_o) \\ f(x_1,y_o) \\ f(x_2,y_o) \end{bmatrix} \right\rangle - \left\langle V\left(\frac{\delta_x^U}{h_x}\right), \begin{bmatrix} f(x_{n_x},y_o) \\ f(x_{n_x-1},y_o) \\ f(x_{n_x-2},y_o) \end{bmatrix} \right\rangle \end{array} \right] \\
+4 \sum_{j=0}^{\frac{n_y}{2}-1} \frac{h_x}{3} \left[\begin{array}{l} f(x_o,y_{2j+1}) + 4 \sum_{i=0}^{\frac{n_x}{2}-1} f(x_{2i+1},y_{2j+1}) + 2 \sum_{i=1}^{\frac{n_x}{2}-1} f(x_{2i},y_{2j+1}) + f(x_{n_x},y_{2j+1}) \\ - \left\langle V\left(\frac{\delta_x^L}{h_x}\right), \begin{bmatrix} f(x_o,y_{2j+1}) \\ f(x_1,y_{2j+1}) \\ f(x_2,y_{2j+1}) \end{bmatrix} \right\rangle - \left\langle V\left(\frac{\delta_x^U}{h_x}\right), \begin{bmatrix} f(x_{n_x},y_{2j+1}) \\ f(x_{n_x-1},y_{2j+1}) \\ f(x_{n_x-2},y_{2j+1}) \end{bmatrix} \right\rangle \end{array} \right] \\
+2 \sum_{j=1}^{\frac{n_y}{2}-1} \frac{h_x}{3} \left[\begin{array}{l} f(x_o,y_{2j}) + 4 \sum_{i=0}^{\frac{n_x}{2}-1} f(x_{2i+1},y_{2j}) + 2 \sum_{i=1}^{\frac{n_x}{2}-1} f(x_{2i},y_{2j}) + f(x_{n_x},y_{2j}) \\ - \left\langle V\left(\frac{\delta_x^L}{h_x}\right), \begin{bmatrix} f(x_o,y_{2j}) \\ f(x_1,y_{2j}) \\ f(x_2,y_{2j}) \end{bmatrix} \right\rangle - \left\langle V\left(\frac{\delta_x^U}{h_x}\right), \begin{bmatrix} f(x_{n_x},y_{2j}) \\ f(x_{n_x-1},y_{2j}) \\ f(x_{n_x-2},y_{2j}) \end{bmatrix} \right\rangle \end{array} \right] \\
\frac{h_x}{3} \left[\begin{array}{l} f(x_o,y_{n_y}) + 4 \sum_{i=0}^{\frac{n_x}{2}-1} f(x_{2i+1},y_{n_y}) + 2 \sum_{i=1}^{\frac{n_x}{2}-1} f(x_{2i},y_{n_y}) + f(x_{n_x},y_{n_y}) \\ - \left\langle V\left(\frac{\delta_x^L}{h_x}\right), \begin{bmatrix} f(x_o,y_{n_x}) \\ f(x_1,y_{n_x}) \\ f(x_2,y_{n_x}) \end{bmatrix} \right\rangle - \left\langle V\left(\frac{\delta_x^U}{h_x}\right), \begin{bmatrix} f(x_{n_x},y_{n_x}) \\ f(x_{n_x-1},y_{n_x}) \\ f(x_{n_x-2},y_{n_x}) \end{bmatrix} \right\rangle \end{array} \right]
\end{array} \right\}$$

$$
-\frac{h_y}{3}\times
$$

$$
\times\left\{
\begin{aligned}
&\left\langle V\left(\frac{\delta_y^L}{h_y}\right),
\begin{bmatrix}
\frac{h_x}{3}\left[f(x_o,y_o) + 4\sum_{i=0}^{\frac{n_x}{2}-1} f(x_{2i+1},y_o) + 2\sum_{i=1}^{\frac{n_x}{2}-1} f(x_{2i},y_o) + f(x_{n_x},y_o) - \left\langle V\left(\frac{\delta_x^L}{h_x}\right), \begin{bmatrix} f(x_o,y_o)\\ f(x_1,y_o)\\ f(x_2,y_o)\end{bmatrix}\right\rangle - \left\langle V\left(\frac{\delta_x^U}{h_x}\right), \begin{bmatrix} f(x_{n_x},y_o)\\ f(x_{n_x-1},y_o)\\ f(x_{n_x-2},y_o)\end{bmatrix}\right\rangle \right] \\
\frac{h_x}{3}\left[f(x_o,y_1) + 4\sum_{i=0}^{\frac{n_x}{2}-1} f(x_{2i+1},y_1) + 2\sum_{i=1}^{\frac{n_x}{2}-1} f(x_{2i},y_1) + f(x_{n_x},y_1) - \left\langle V\left(\frac{\delta_x^L}{h_x}\right), \begin{bmatrix} f(x_o,y_1)\\ f(x_1,y_1)\\ f(x_2,y_1)\end{bmatrix}\right\rangle - \left\langle V\left(\frac{\delta_x^U}{h_x}\right), \begin{bmatrix} f(x_{n_x},y_1)\\ f(x_{n_x-1},y_1)\\ f(x_{n_x-2},y_1)\end{bmatrix}\right\rangle \right] \\
\frac{h_x}{3}\left[f(x_o,y_2) + 4\sum_{i=0}^{\frac{n_x}{2}-1} f(x_{2i+1},y_2) + 2\sum_{i=1}^{\frac{n_x}{2}-1} f(x_{2i},y_2) + f(x_{n_x},y_2) - \left\langle V\left(\frac{\delta_x^L}{h_x}\right), \begin{bmatrix} f(x_o,y_2)\\ f(x_1,y_2)\\ f(x_2,y_2)\end{bmatrix}\right\rangle - \left\langle V\left(\frac{\delta_x^U}{h_x}\right), \begin{bmatrix} f(x_{n_x},y_2)\\ f(x_{n_x-1},y_2)\\ f(x_{n_x-2},y_2)\end{bmatrix}\right\rangle \right]
\end{bmatrix}\right\rangle \\
&+\left\langle V\left(\frac{\delta_y^U}{h_y}\right),
\begin{bmatrix}
\frac{h_x}{3}\left[f(x_o,y_{n_y}) + 4\sum_{i=0}^{\frac{n_x}{2}-1} f(x_{2i+1},y_{n_y}) + 2\sum_{i=1}^{\frac{n_x}{2}-1} f(x_{2i},y_{n_y}) + f(x_{n_x},y_{n_y}) - \left\langle V\left(\frac{\delta_x^L}{h_x}\right), \begin{bmatrix} f(x_o,y_{n_y})\\ f(x_1,y_{n_y})\\ f(x_2,y_{n_y})\end{bmatrix}\right\rangle - \left\langle V\left(\frac{\delta_x^U}{h_x}\right), \begin{bmatrix} f(x_{n_x},y_{n_y})\\ f(x_{n_x-1},y_{n_y})\\ f(x_{n_x-2},y_{n_y})\end{bmatrix}\right\rangle \right] \\
\frac{h_x}{3}\left[f(x_o,y_{n_y-1}) + 4\sum_{i=0}^{\frac{n_x}{2}-1} f(x_{2i+1},y_{n_y-1}) + 2\sum_{i=1}^{\frac{n_x}{2}-1} f(x_{2i},y_{n_y-1}) + f(x_{n_x},y_{n_y-1}) - \left\langle V\left(\frac{\delta_x^L}{h_x}\right), \begin{bmatrix} f(x_o,y_{n_y-1})\\ f(x_1,y_{n_y-1})\\ f(x_2,y_{n_y-1})\end{bmatrix}\right\rangle - \left\langle V\left(\frac{\delta_x^U}{h_x}\right), \begin{bmatrix} f(x_{n_x},y_{n_y-1})\\ f(x_{n_x-1},y_{n_y-1})\\ f(x_{n_x-2},y_{n_y-1})\end{bmatrix}\right\rangle \right] \\
\frac{h_x}{3}\left[f(x_o,y_{n_y-2}) + 4\sum_{i=0}^{\frac{n_x}{2}-1} f(x_{2i+1},y_{n_y-2}) + 2\sum_{i=1}^{\frac{n_x}{2}-1} f(x_{2i},y_{n_y-2}) + f(x_{n_x},y_{n_y-2}) - \left\langle V\left(\frac{\delta_x^L}{h_x}\right), \begin{bmatrix} f(x_o,y_{n_y-2})\\ f(x_1,y_{n_y-2})\\ f(x_2,y_{n_y-2})\end{bmatrix}\right\rangle - \left\langle V\left(\frac{\delta_x^U}{h_x}\right), \begin{bmatrix} f(x_{n_x},y_{n_y-2})\\ f(x_{n_x-1},y_{n_y-2})\\ f(x_{n_x-2},y_{n_y-2})\end{bmatrix}\right\rangle \right]
\end{bmatrix}\right\rangle
\end{aligned}
\right\}, \tag{5.20}
$$

where

$$
h_y = \frac{y_{n_y} - y_o}{n_y} = y_{i+1} - y_i\,, \qquad i = (0, \ldots, n_y - 1);\ n_y \text{ even}\,. \tag{5.21}
$$

Simplifying Equation 5.20, we obtain:

$$
\int_{y_o+\delta_y^L}^{y_{n_y}-\delta_y^U} \int_{x_o+\delta_x^L}^{x_{n_x}-\delta_x^U} f(x,y)\,dx\,dy \approx
$$

$$\frac{h_x h_y}{9}\left\{\begin{aligned}&f(x_o,y_o)+f(x_o,y_{n_y})+f(x_{n_x},y_o)+f(x_{n_x},y_{n_y})\\&+2\left[\sum_{i=1}^{\frac{n_x}{2}-1}\Big(f(x_{2i},y_o)+f(x_{2i},y_{n_y})\Big)+\sum_{j=1}^{\frac{n_y}{2}-1}\Big(f(x_o,y_{2j})+f(x_{n_x},y_{2j})\Big)\right]\\&+4\left[\sum_{i=0}^{\frac{n_x}{2}-1}\Big(f(x_{2i+1},y_o)+f(x_{2i+1},y_{n_y})\Big)+\sum_{j=0}^{\frac{n_y}{2}-1}\Big(f(x_o,y_{2j+1})+f(x_{n_x},y_{2j+1})\Big)\right]\\&+8\left(\sum_{j=0}^{\frac{n_y}{2}-1}\sum_{i=1}^{\frac{n_x}{2}-1}f(x_{2i},y_{2j+1})+\sum_{j=1}^{\frac{n_y}{2}-1}\sum_{i=0}^{\frac{n_x}{2}-1}f(x_{2i+1},y_{2j})\right)\\&+4\sum_{j=1}^{\frac{n_y}{2}-1}\sum_{i=1}^{\frac{n_x}{2}-1}f(x_{2i},y_{2j})+16\sum_{j=0}^{\frac{n_y}{2}-1}\sum_{i=0}^{\frac{n_x}{2}-1}f(x_{2i+1},y_{2j+1})\end{aligned}\right\}$$

$$-\frac{h_x h_y}{9}\left\{\begin{aligned}&\left\langle V\left(\frac{\delta_x^L}{h_x}\right),\begin{bmatrix}f(x_o,y_o)\\f(x_1,y_o)\\f(x_2,y_o)\end{bmatrix}\right\rangle+\left\langle V\left(\frac{\delta_x^U}{h_x}\right),\begin{bmatrix}f(x_{n_x},y_o)\\f(x_{n_x-1},y_o)\\f(x_{n_x-2},y_o)\end{bmatrix}\right\rangle\\&+4\sum_{j=0}^{\frac{n_y}{2}-1}\left[\left\langle V\left(\frac{\delta_x^L}{h_x}\right),\begin{bmatrix}f(x_o,y_{2j+1})\\f(x_1,y_{2j+1})\\f(x_2,y_{2j+1})\end{bmatrix}\right\rangle+\left\langle V\left(\frac{\delta_x^U}{h_x}\right),\begin{bmatrix}f(x_{n_x},y_{2j+1})\\f(x_{n_x-1},y_{2j+1})\\f(x_{n_x-2},y_{2j+1})\end{bmatrix}\right\rangle\right]\\&+2\sum_{j=1}^{\frac{n_y}{2}-1}\left[\left\langle V\left(\frac{\delta_x^L}{h_x}\right),\begin{bmatrix}f(x_o,y_{2j})\\f(x_1,y_{2j})\\f(x_2,y_{2j})\end{bmatrix}\right\rangle+\left\langle V\left(\frac{\delta_x^U}{h_x}\right),\begin{bmatrix}f(x_{n_x},y_{2j})\\f(x_{n_x-1},y_{2j})\\f(x_{n_x-2},y_{2j})\end{bmatrix}\right\rangle\right]\\&+\left\langle V\left(\frac{\delta_x^L}{h_x}\right),\begin{bmatrix}f(x_o,y_{n_y})\\f(x_1,y_{n_y})\\f(x_2,y_{n_y})\end{bmatrix}\right\rangle+\left\langle V\left(\frac{\delta_x^U}{h_x}\right),\begin{bmatrix}f(x_{n_x},y_{n_y})\\f(x_{n_x-1},y_{n_y})\\f(x_{n_x-2},y_{n_y})\end{bmatrix}\right\rangle\end{aligned}\right\}$$

$$
-\frac{h_x h_y}{9}\left\{\begin{array}{l}
\left\langle V\left(\frac{\delta_y^L}{h_y}\right), \begin{bmatrix} f(x_o,y_o)\\ f(x_o,y_1)\\ f(x_o,y_2)\end{bmatrix}\right\rangle + \left\langle V\left(\frac{\delta_y^U}{h_y}\right), \begin{bmatrix} f(x_o,y_{n_y})\\ f(x_o,y_{n_y-1})\\ f(x_o,y_{n_y-2})\end{bmatrix}\right\rangle \\[2ex]
+4\sum_{i=0}^{\frac{n_x}{2}-1}\left[\left\langle V\left(\frac{\delta_y^L}{h_y}\right), \begin{bmatrix} f(x_{2i+1},y_o)\\ f(x_{2i+1},y_1)\\ f(x_{2i+1},y_2)\end{bmatrix}\right\rangle + \left\langle V\left(\frac{\delta_y^U}{h_y}\right), \begin{bmatrix} f(x_{2i+1},y_{n_y})\\ f(x_{2i+1},y_{n_y-1})\\ f(x_{2i+1},y_{n_y-2})\end{bmatrix}\right\rangle\right] \\[2ex]
+2\sum_{i=1}^{\frac{n_x}{2}-1}\left[\left\langle V\left(\frac{\delta_y^L}{h_y}\right), \begin{bmatrix} f(x_{2i},y_o)\\ f(x_{2i},y_1)\\ f(x_{2i},y_2)\end{bmatrix}\right\rangle + \left\langle V\left(\frac{\delta_y^U}{h_y}\right), \begin{bmatrix} f(x_{2i},y_{n_y})\\ f(x_{2i},y_{n_y-1})\\ f(x_{2i},y_{n_y-2})\end{bmatrix}\right\rangle\right] \\[2ex]
+\left\langle V\left(\frac{\delta_y^L}{h_y}\right), \begin{bmatrix} f(x_{n_x},y_o)\\ f(x_{n_x},y_1)\\ f(x_{n_x},y_2)\end{bmatrix}\right\rangle + \left\langle V\left(\frac{\delta_y^U}{h_y}\right), \begin{bmatrix} f(x_{n_x},y_{n_y})\\ f(x_{n_x},y_{n_y-1})\\ f(x_{n_x},y_{n_y-2})\end{bmatrix}\right\rangle
\end{array}\right\}
$$

$$
+\frac{h_x h_y}{9}\left\{\begin{array}{l}
\left\langle V\left(\frac{\delta_y^L}{h_y}\right), \begin{bmatrix}
\left\langle V\left(\frac{\delta_x^L}{h_x}\right), \begin{bmatrix} f(x_o,y_o)\\ f(x_1,y_o)\\ f(x_2,y_o)\end{bmatrix}\right\rangle \\
\left\langle V\left(\frac{\delta_x^L}{h_x}\right), \begin{bmatrix} f(x_o,y_1)\\ f(x_1,y_1)\\ f(x_2,y_1)\end{bmatrix}\right\rangle \\
\left\langle V\left(\frac{\delta_x^L}{h_x}\right), \begin{bmatrix} f(x_o,y_2)\\ f(x_1,y_2)\\ f(x_2,y_2)\end{bmatrix}\right\rangle
\end{bmatrix}\right\rangle + \left\langle V\left(\frac{\delta_y^L}{h_y}\right), \begin{bmatrix}
\left\langle V\left(\frac{\delta_x^U}{h_x}\right), \begin{bmatrix} f(x_{n_x},y_o)\\ f(x_{n_x-1},y_o)\\ f(x_{n_x-2},y_o)\end{bmatrix}\right\rangle \\
\left\langle V\left(\frac{\delta_x^U}{h_x}\right), \begin{bmatrix} f(x_{n_x},y_1)\\ f(x_{n_x-1},y_1)\\ f(x_{n_x-2},y_1)\end{bmatrix}\right\rangle \\
\left\langle V\left(\frac{\delta_x^U}{h_x}\right), \begin{bmatrix} f(x_{n_x},y_2)\\ f(x_{n_x-1},y_2)\\ f(x_{n_x-2},y_2)\end{bmatrix}\right\rangle
\end{bmatrix}\right\rangle \\[3ex]
+\left\langle V\left(\frac{\delta_y^U}{h_y}\right), \begin{bmatrix}
\left\langle V\left(\frac{\delta_x^L}{h_x}\right), \begin{bmatrix} f(x_o,y_{n_y})\\ f(x_1,y_{n_y})\\ f(x_2,y_{n_y})\end{bmatrix}\right\rangle \\
\left\langle V\left(\frac{\delta_x^L}{h_x}\right), \begin{bmatrix} f(x_o,y_{n_y-1})\\ f(x_1,y_{n_y-1})\\ f(x_2,y_{n_y-1})\end{bmatrix}\right\rangle \\
\left\langle V\left(\frac{\delta_x^L}{h_x}\right), \begin{bmatrix} f(x_o,y_{n_y-2})\\ f(x_1,y_{n_y-2})\\ f(x_2,y_{n_y-2})\end{bmatrix}\right\rangle
\end{bmatrix}\right\rangle + \left\langle V\left(\frac{\delta_y^U}{h_y}\right), \begin{bmatrix}
\left\langle V\left(\frac{\delta_x^U}{h_x}\right), \begin{bmatrix} f(x_{n_x},y_{n_y})\\ f(x_{n_x-1},y_{n_y})\\ f(x_{n_x-2},y_{n_y})\end{bmatrix}\right\rangle \\
\left\langle V\left(\frac{\delta_x^U}{h_x}\right), \begin{bmatrix} f(x_{n_x},y_{n_y-1})\\ f(x_{n_x-1},y_{n_y-1})\\ f(x_{n_x-2},y_{n_y-1})\end{bmatrix}\right\rangle \\
\left\langle V\left(\frac{\delta_x^U}{h_x}\right), \begin{bmatrix} f(x_{n_x},y_{n_y-2})\\ f(x_{n_x-1},y_{n_y-2})\\ f(x_{n_x-2},y_{n_y-2})\end{bmatrix}\right\rangle
\end{bmatrix}\right\rangle
\end{array}\right\}. \tag{5.22}
$$

Performing the summation over each sample of $f(x,y)$, Equation 5.22 can be expressed in a more compact nomenclature:

$$
\int_{y_o+\delta_y^L}^{y_f-\delta_y^U}\int_{x_o+\delta_x^L}^{x_f-\delta_x^U} f(x,y)\,dx\,dy
$$

$$
=\frac{h_x h_y}{9}\sum_{i=0}^{n_x}\sum_{j=0}^{n_y}\left(S_{i,j}-R_{i,j}\left(h_x,h_y,\delta_x^L,\delta_x^H,\delta_y^L,\delta_y^H\right)\right)f(x_i,y_j)+\varepsilon(f)\,, \tag{5.23}
$$

where

$$
R\left(h_x,h_y,\delta_x^L,\delta_x^H,\delta_y^L,\delta_y^H\right) =
\begin{bmatrix}
V_1^{xL}+V_1^{yL} & V_2^{xL}+4V_1^{yL} & V_3^{xL}+2V_1^{yL} & 4V_1^{yL} & 2V_1^{yL} & 4V_1^{yL} & 2V_1^{yL} & \cdots & 4V_1^{yL} & 2V_1^{yL} & 4V_1^{yL} & V_3^{xU}+2V_1^{yL} & V_2^{xU}+4V_1^{yL} & V_1^{xU}+V_1^{yL} \\
4V_1^{xL}+V_2^{yL} & 4V_2^{xL}+4V_2^{yL} & 4V_3^{xL}+2V_2^{yL} & 4V_2^{yL} & 2V_2^{yL} & 4V_2^{yL} & 2V_2^{yL} & \cdots & 4V_2^{yL} & 2V_2^{yL} & 4V_2^{yL} & 4V_3^{xU}+2V_2^{yL} & 4V_2^{xU}+4V_2^{yL} & 4V_1^{xU}+V_2^{yL} \\
2V_1^{xL}+V_3^{yL} & 2V_2^{xL}+4V_3^{yL} & 2V_3^{xL}+2V_3^{yL} & 4V_3^{yL} & 2V_3^{yL} & 4V_3^{yL} & 2V_3^{yL} & \cdots & 4V_3^{yL} & 2V_3^{yL} & 4V_3^{yL} & 2V_3^{xU}+2V_3^{yL} & 2V_2^{xU}+4V_3^{yL} & 2V_1^{xU}+V_3^{yL} \\
4V_1^{xL} & 4V_2^{xL} & 4V_3^{xL} & 0 & 0 & 0 & 0 & \cdots & 0 & 0 & 0 & 4V_3^{xU} & 4V_2^{xU} & 4V_1^{xU} \\
2V_1^{xL} & 2V_2^{xL} & 2V_3^{xL} & 0 & \ddots & & & & & \iddots & 0 & 2V_3^{xU} & 2V_2^{xU} & 2V_1^{xU} \\
4V_1^{xL} & 4V_2^{xL} & 4V_3^{xL} & 0 & & & & & & & 0 & 4V_3^{xU} & 4V_2^{xU} & 4V_1^{xU} \\
2V_1^{xL} & 2V_2^{xL} & 2V_3^{xL} & 0 & & & & & & & 0 & 2V_3^{xU} & 2V_2^{xU} & 2V_1^{xU} \\
\vdots & \vdots & \vdots & \vdots & & & & & & & \vdots & \vdots & \vdots & \vdots \\
4V_1^{xL} & 4V_2^{xL} & 4V_3^{xL} & 0 & & & & & & & 0 & 4V_3^{xU} & 4V_2^{xU} & 4V_1^{xU} \\
2V_1^{xL} & 2V_2^{xL} & 2V_3^{xL} & 0 & \iddots & & & & & \ddots & 0 & 2V_3^{xU} & 2V_2^{xU} & 2V_1^{xU} \\
4V_1^{xL} & 4V_2^{xL} & 4V_3^{xL} & 0 & 0 & 0 & 0 & \cdots & 0 & 0 & 0 & 4V_3^{xU} & 4V_2^{xU} & 4V_1^{xU} \\
2V_1^{xL}+V_3^{yU} & 2V_2^{xL}+4V_3^{yU} & 2V_3^{xL}+2V_3^{yU} & 4V_3^{yU} & 2V_3^{yU} & 4V_3^{yU} & 2V_3^{yU} & \cdots & 4V_3^{yU} & 2V_3^{yU} & 4V_3^{yU} & 2V_3^{xU}+2V_3^{yU} & 2V_2^{xU}+4V_3^{yU} & 2V_1^{xU}+V_3^{yU} \\
4V_1^{xL}+V_2^{yU} & 4V_2^{xL}+4V_2^{yU} & 4V_3^{xL}+2V_2^{yU} & 4V_2^{yU} & 2V_2^{yU} & 4V_2^{yU} & 2V_2^{yU} & \cdots & 4V_2^{yU} & 2V_2^{yU} & 4V_2^{yU} & 4V_3^{xU}+2V_2^{yU} & 4V_2^{xU}+4V_2^{yU} & 4V_1^{xU}+V_2^{yU} \\
V_1^{xL}+V_1^{yU} & V_2^{xL}+4V_1^{yU} & V_3^{xL}+2V_1^{yU} & 4V_1^{yU} & 2V_1^{yU} & 4V_1^{yU} & 2V_1^{yU} & \cdots & 4V_1^{yU} & 2V_1^{yU} & 4V_1^{yU} & V_3^{xU}+2V_1^{yU} & V_2^{xU}+4V_1^{yU} & V_1^{xU}+V_1^{yU}
\end{bmatrix}
$$

$$
-\begin{bmatrix}
V_1^{xL}V_1^{yL} & V_2^{xL}V_1^{yL} & V_3^{xL}V_1^{yL} & 0 & \cdots & \cdots & 0 & V_3^{xU}V_1^{yL} & V_2^{xU}V_1^{yL} & V_1^{xU}V_1^{yL} \\
V_1^{xL}V_2^{yL} & V_2^{xL}V_2^{yL} & V_3^{xL}V_2^{yL} & & & & & V_3^{xU}V_2^{yL} & V_2^{xU}V_2^{yL} & V_1^{xU}V_2^{yL} \\
V_1^{xL}V_3^{yL} & V_2^{xL}V_3^{yL} & V_3^{xL}V_3^{yL} & & & & & V_3^{xU}V_3^{yL} & V_2^{xU}V_3^{yL} & V_1^{xU}V_3^{yL} \\
0 & 0 & 0 & 0 & \cdots & \cdots & 0 & 0 & 0 & 0 \\
\vdots & \vdots & \vdots & & \ddots & \iddots & & \vdots & \vdots & \vdots \\
\vdots & \vdots & \vdots & & \iddots & \ddots & & \vdots & \vdots & \vdots \\
0 & 0 & 0 & 0 & & & 0 & 0 & 0 & 0 \\
V_1^{xL}V_3^{yU} & V_2^{xL}V_3^{yU} & V_3^{xL}V_3^{yU} & 0 & \cdots & \cdots & 0 & V_3^{xU}V_3^{yU} & V_2^{xU}V_3^{yU} & V_1^{xU}V_3^{yU} \\
V_1^{xL}V_2^{yU} & V_2^{xL}V_2^{yU} & V_3^{xL}V_2^{yU} & & & & & V_3^{xU}V_2^{yU} & V_2^{xU}V_2^{yU} & V_1^{xU}V_2^{yU} \\
V_1^{xL}V_1^{yU} & V_2^{xL}V_1^{yU} & V_3^{xL}V_1^{yU} & 0 & \cdots & \cdots & 0 & V_3^{xU}V_1^{yU} & V_2^{xU}V_1^{yU} & V_1^{xU}V_1^{yU}
\end{bmatrix}, \quad (5.24)
$$

where

V_i^{xL} is the ith component of the vector $V\left(\frac{\delta_x^L}{h_x}\right)$

V_i^{xU} is the ith component of the vector $V\left(\frac{\delta_x^U}{h_x}\right)$

V_i^{yL} is the ith component of the vector $V\left(\frac{\delta_y^L}{h_y}\right)$

V_i^{yU} is the ith component of the vector $V\left(\frac{\delta_y^U}{h_y}\right)$

$$S = \overbrace{\begin{bmatrix} 1 & 4 & 2 & 4 & 2 & \cdots & 4 & 2 & 4 & 1 \\ 4 & 16 & 8 & 16 & 8 & \cdots & 16 & 8 & 16 & 4 \\ 2 & 8 & 4 & 8 & 4 & \cdots & 8 & 4 & 8 & 2 \\ 4 & 16 & 8 & 16 & 8 & \cdots & 16 & 8 & 16 & 4 \\ 2 & 8 & 4 & 8 & 4 & \cdots & 8 & 4 & 8 & 2 \\ \vdots & \vdots & \vdots & \vdots & \vdots & & \vdots & \vdots & \vdots & \vdots \\ 4 & 16 & 8 & 16 & 8 & \cdots & 16 & 8 & 16 & 4 \\ 2 & 8 & 4 & 8 & 4 & \cdots & 8 & 4 & 8 & 2 \\ 4 & 16 & 8 & 16 & 8 & \cdots & 16 & 8 & 16 & 4 \\ 1 & 4 & 2 & 4 & 2 & \cdots & 4 & 2 & 4 & 1 \end{bmatrix}}^{n_x + 1} \left.\vphantom{\begin{bmatrix} 1 \\ 1 \\ 1 \\ 1 \\ 1 \\ 1 \\ 1 \\ 1 \\ 1 \\ 1 \end{bmatrix}}\right\} n_y + 1 \,, \tag{5.25}$$

and $\varepsilon(f)$ is the integration error due to the numerical approximation process.

Equation 5.23 reveals that knowledge of the 2-D function $f(x, y)$ at the knots of a rectangular evenly spaced grid allows noncollocated numerical integration with an approximation error $\varepsilon(f)$. The error, $\varepsilon(f)$, is evaluated in Appendix C, where it is shown that the integration error depends on the spacing of the high-resolution sample grid, *i.e.*, h_x, h_y, and can be made arbitrarily small by choosing these grid parameters appropriately. The way to derive a suitable grid spacing h_x, h_y from a specified maximum error is also presented in Appendix C.

6 Computational Tools

A computational tool, called EyeBox, was created to model and visualize the extensive image formation process, as well as to evaluate the CRLB and practical estimators. This tool was developed using MATLAB®. EyeBox provides a framework within which the parameters of the physical model of image formation are organized, enabling the investigation of landmark location uncertainty using the analytical framework presented in Chapter 3.

A block diagram of the EyeBox tool is shown in Figure 6.1. This diagram also shows the high-level structure of the experimental tool EyeExpBox, which will be introduced in Chapter 7.

The functional architecture of EyeBox is shown in Plate 6 (see Color Section, page 149). The main inputs to the tool are the parameters provided by a configuration file and an experiment name. From here, a sequence of components implements the image formation transformations indicated in Equation 2.4, to estimate landmark position, based on the user-selected estimators, and the related confidence intervals. Table 6.1 presents an example of the typical configuration file used in EyeBox.

In this chapter, complete descriptions of the EyeBox components are presented, including details on the mathematical treatment of the image formation processes and the calculation of the landmark location uncertainty via the confidence intervals generated.

6.1 EyeEntity

EyeEntity is the first component in EyeBox. EyeEntity creates a high-resolution numerical model of the optical flux function $H(x,y)$ of a circular landmark. $H(x,y)$ is created via the pinhole model associated with geometric optics, using the geometrical parameters listed in the configuration file. These parameters are the landmark radius and the six parameters describing the landmark pose in camera coordinates. The sampling density of the high-resolution grid is also provided by the user in the configuration file; this parameter should be calculated according to the 2-D integration error requirements of the spatial sampling presented in Chapter 5 and Appendix C.

EyeEntity also creates the illuminance function $E(x,y)$. This is achieved by convolving the optical flux function $H(x,y)$ with the 2-D impulse response of the

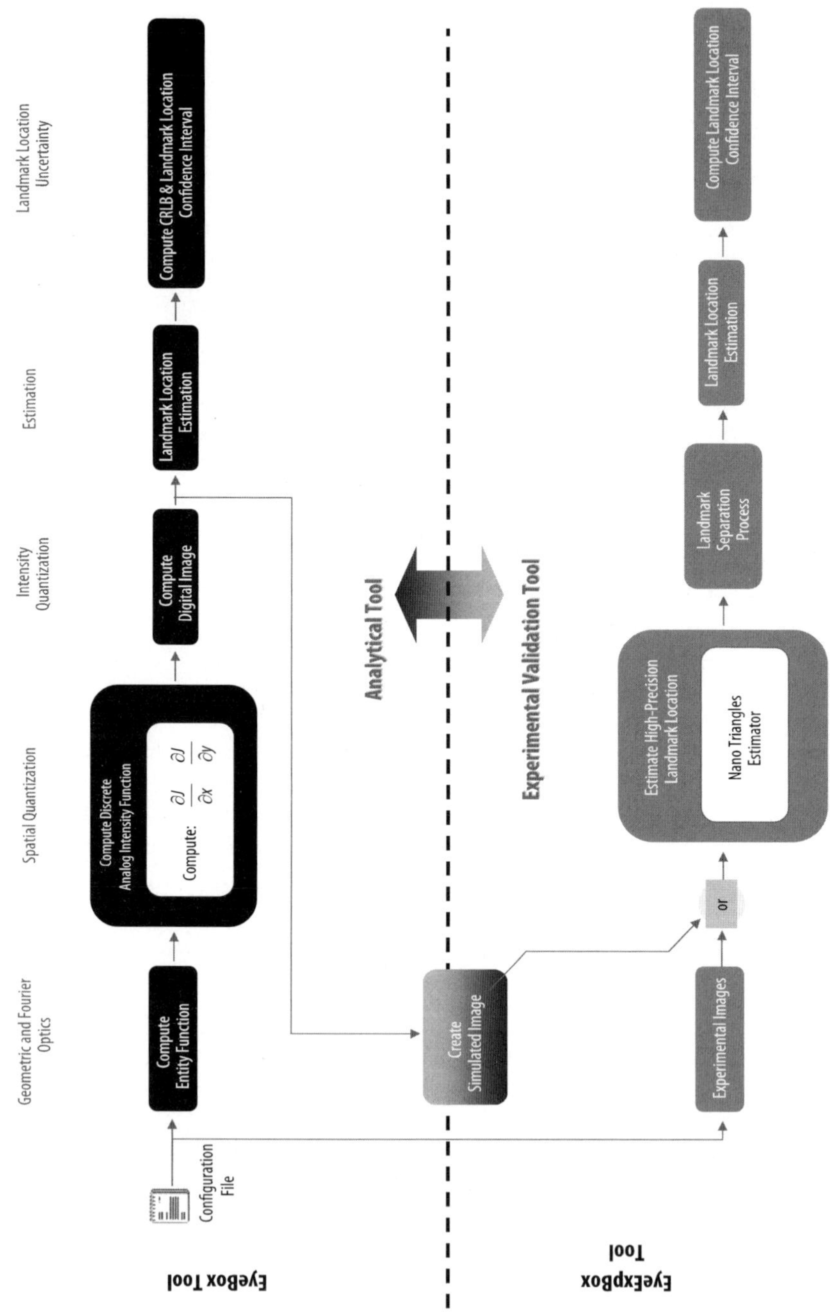

Figure 6.1. Block diagram of EyeBox and EyeExpBox

Table 6.1. Example configuration file for EyeBox. Types are 's' string, 'd' integer, 'f' floating point, and 'v' vector of floating point.

	Parameter	Value	Type	Description
Camera	CameraName	PULNiX9701	s	Imager mode based on the Pulnix 9701
	LensName	35mmDG	s	Double Gauss 35-mm lens
	FrameGrabberName	mv1500	s	Simulated
	N_x	20	d	Imager resolution: horizontal [pixels]
	N_y	20	d	Imager resolution: vertical [pixels]
	Sblur	0.005	f	Diffraction blur radius [mm]
	Std. Deviation Noise	0.004	f	Standard deviation of noise in imager: $1/2^8$
	w_x	80%	f	Percentage pixel sensitive area on x
	w_y	80%	f	Percentage pixel sensitive area on y
	k_x	83	f	Pixel density in x. $k_x = N_x$/width [pixel/mm]
	k_y	73	f	Pixel density in y. $k_y = N_y$/height [pixel/mm]
	x_0	10	f	Principal point, x coordinate [pixels]
	y_0	10	f	Principal point, y coordinate [pixels]
	c_p	35	f	Principal distance [mm]
	K	[0 0]	v	Radial lens distortion parameters
	P	[0 0]	v	Decentering lens distortion parameters
	bZoomLens	0	f	No zoom
Land-mark	ConfigName	TestDisk	s	Small disk
	ArtworkName	Disk	s	Circular landmark
	cR_L	3	f	Landmark radius [mm]
	cX_L	0	f	x coordinate of landmark center in camera coordinates [mm]
	cY_L	0	f	y coordinate of landmark center in camera coordinates [mm]
	cZ_L	2700	f	z coordinate of landmark center in camera coordinates [mm]
	cPitch_L	0	f	Pitch of target in camera coordinates
	cRoll_L	10	f	Roll of target in camera coordinates
	cYaw_L	0	f	Yaw of target in camera coordinates
	Lmin	0.2	f	Luminosity dark [fraction of full scale]
	Lmax	0.8	f	Luminosity light [fraction of full scale]
	Rect	[]	v	For use in EyeExpBox
	KIG	0	f	Nonuniform illumination rate [intensity/mm]
	AngleIG	0	f	Nonuniform illumination plane angle [degrees]

lens system. The spatial impulse response is approximated by a 2-D Gaussian with variance σ_b^2 (see Section 2.1), where σ_b is the diffraction smoothing parameter from the configuration file (see Plate 7 in the Color Section, page 150). The Gaussian kernel used in the smoothing process is sampled with the same spatial grid used in the creation of $H(x, y)$. A sample screenshot of the EyeEntity component is shown in Figure 6.2.

EyeEntity stores landmark, camera, and other ancillary description information, along with the generated optical flux and illuminance functions, in a MATLAB® data file for subsequent processing.

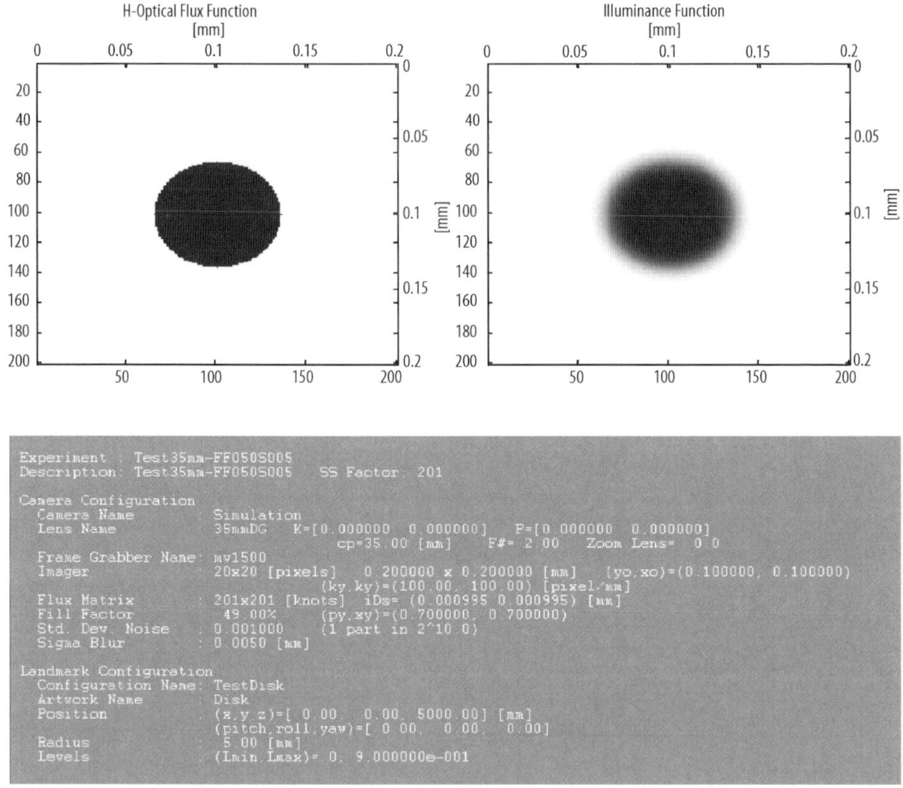

Figure 6.2. Example screenshot of EyeEntity

6.2 EyeQs and EyeRasterQs

EyeQs performs the spatial sampling process on the illuminance function $E(x,y)$ previously generated by EyeEntity. EyeQs uses the resolution (N_x, N_y), the pixel density (k_x, k_y), the principal point (x_0, y_0), and the pixel sensitive area (p_x, p_y) indicated in the configuration file. The spatial sampling is performed using the 2-D noncollocated numerical integration presented in Chapter 5. The spatial sampling process is shown in Figure 6.3.

EyeQs provide the user with the capability to offset the sampling grid with respect to the illuminance function. This feature is particularly relevant to experiments where landmark uncertainty is sampled over a $\pm 1/2$ pixel area.

Once the spatial sampling is performed, user-defined zero-mean Gaussian noise with a variance of σ_n^2 is added to the discrete samples, obtaining the discrete analog intensity function $\bar{J}(i,j) = J^*(i,j) + \widetilde{J}(i,j)$. A sample screenshot of the EyeEntity component is shown in Figure 6.4.

EyeQs also computes the spatial derivatives $\frac{\partial J}{\partial x}$ and $\frac{\partial J}{\partial y}$ required to compute the CRLB and estimator confidence intervals. The derivatives of the area integral $J(j,k)$ are computed by evaluating line integrals at the boundary of the sensitive

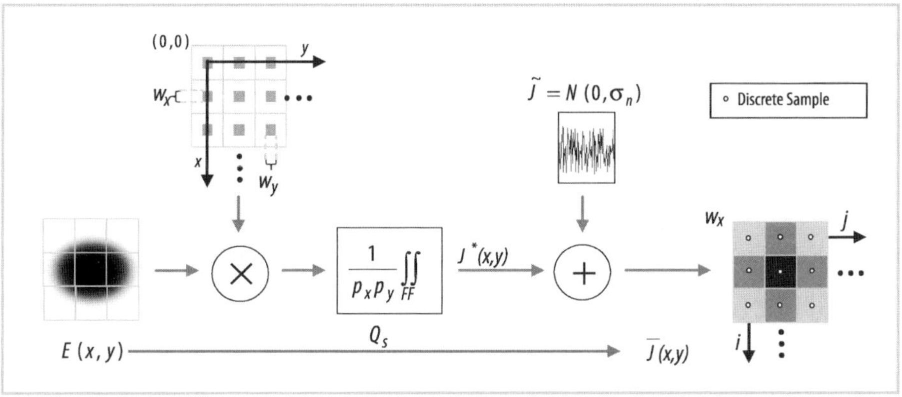

Figure 6.3. Spatial sampling process

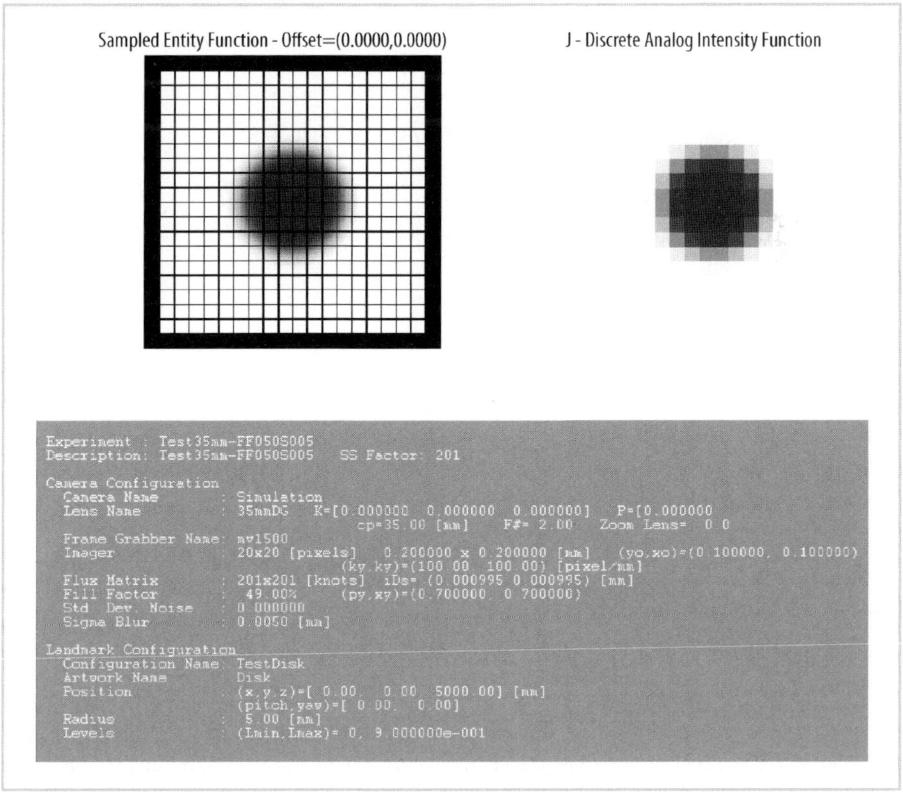

Figure 6.4. Example screenshot of EyeQs

area corresponding to $J(j,k)$. These line integrations are performed within the 2-D noncollocated numerical integration routine. Numerical first difference is not used.

EyeRasterQs is functionally similar to EyeQs, but was created to perform automated raster spatial quantization in the $\pm 1/2$ pixel area to model the uniform probability distribution of the true landmark location in the pixel. The results from EyeQs and EyeRasterQs are stored for subsequent use of the remaining EyeBox components.

6.3 EyeQi and EyeRasterQi

The amplitude quantization transformation is performed by EyeQi. Using this component, the amplitude of the discrete analog intensity function $\bar{J}(i,j)$ is quantized by a uniform quantizer to form the digital image function $\bar{I}(i,j)$. This process is illustrated in Figure 6.5.

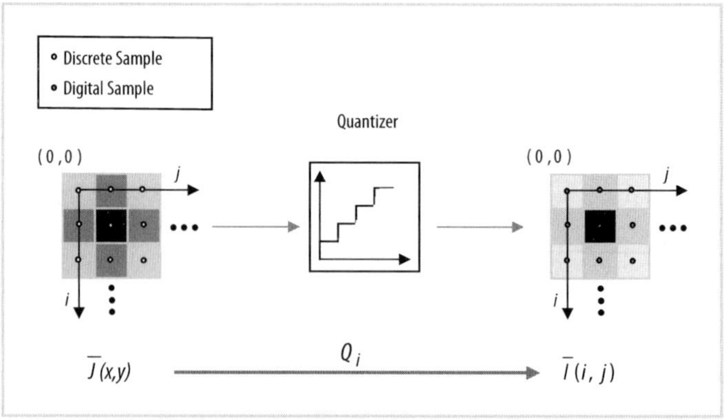

Figure 6.5. Amplitude quantization process

Similarly, EyeRasterQi performs the amplitude quantization task for the discrete analog intensity functions generated by EyeRasterQs.

EyeQi and EyeRasterQi generate digital image functions for selected quantification levels in the range of $q = [1 \ldots 16]$, and the results are stored so that they can be used by the EyeT component described below.

6.4 EyeT

EyeT is the EyeBox component in charge of performing landmark location estimation for both discrete analog amplitude functions and digital images. The estimators used in EyeT are binary centroid, grayscale centroid, ellipsoidal contour, and Butterworth tepuy. In order to accommodate corrections for lens distortion and the geometry of non-square pixels, these estimators are implemented in image coordinates; that is $\left[{}^{i}x,\ {}^{i}y\right]^{T}$, in millimeters, rather than in pixel coordinates. The estimation process is illustrated in Figure 6.6.

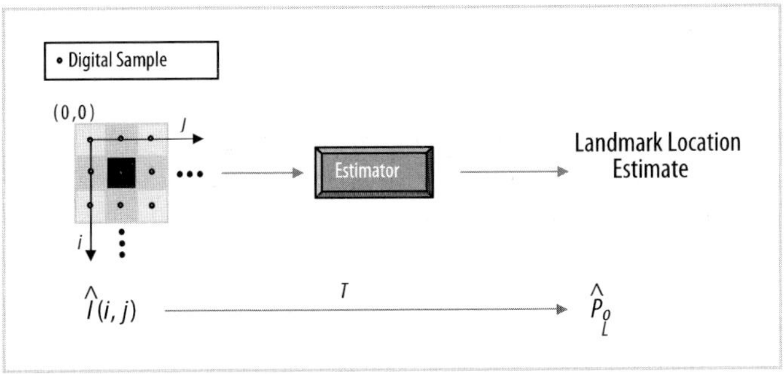

Figure 6.6. Estimation process

EyeT automatically detects whether the experiment corresponds to a single or raster image simulation. In the case of raster simulations, EyeT stores the estimation results in a special folder for further processing by EyeCI.

6.5 EyeCI

The final component in EyeBox is EyeCI. This tool computes the CRLB and estimation confidence intervals according to the framework developed in Chapter 3.

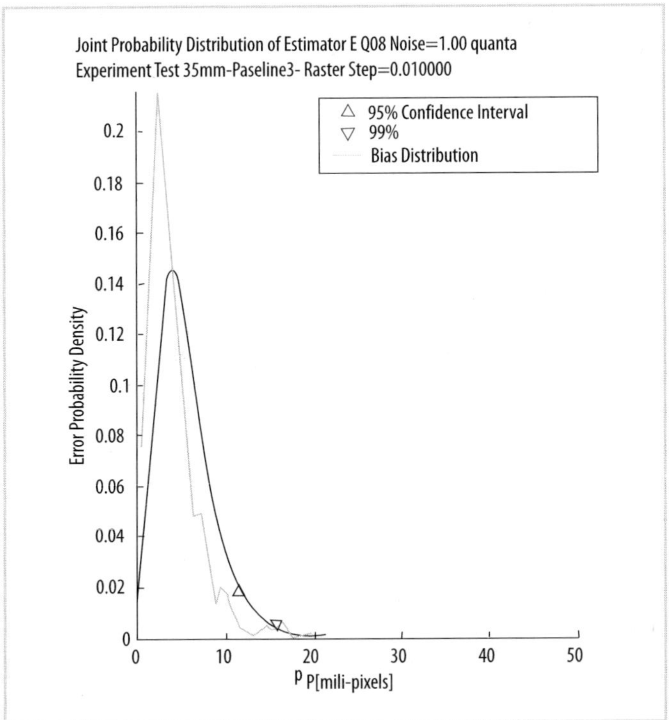

Figure 6.7. Example confidence interval display from EyeCI

Circular integrals from Equations 3.16 and 3.42 are evaluated using regular 2-D Simpson integration (not the noncollocated version introduced in this work). Furthermore, the overall confidence interval integration over the $\pm 1/2$ pixel area, Equations 3.20 and 3.44, was implemented using numerical Euler integration. A sample display of a confidence interval computation performed by EyeCI is shown in Figure 6.7.

The confidence interval shown is the result of processing the configuration called Test35mm-Baseline3 on digital images quantized at eight bits ($q = 8$) using the Butterworth tepuy estimators (denoted "E" in the figure legend). The noise level applied to this experiment was $1/256^{\text{th}}$ of the full scale of the imager considered, modeling $\sigma_b = 1.0$ count for the eight-bit ADC. Figure 6.7 also shows the distribution of the estimator bias; this information is relevant since it provides an indication of the effect of bias on the uncertainty level. For the configuration investigated to produce Plate 7, bias contributes a large portion of the total uncertainty.

Confidence intervals for both the CRLB and practical estimators can be computed with the EyeBox tool. Chapter 8 presents the results from numerical experiments performed using the EyeBox tool in order to study the impact of selected characteristics on precision landmark location.

7 Experimental Validation

This section reports an experiment that was designed to directly measure the uncertainty of landmark location estimates. The experiment is used to test and validate the theoretical framework developed in Chapters 2 and 3. The outcomes of these experiments are confidence interval maps for four landmark location estimators: binary centroid, the grayscale centroid, ellipsoidal contour, and the Butterworth tepuy.

Landmark location estimator error is given by:

$$ {}^{p}\widetilde{P}_{o} = {}^{p}\widehat{P}_{o} - {}^{p}P^{*}_{o} \,, \tag{7.1} $$

where ${}^{p}\widehat{P}_{0}$ is given by the estimator under test. The challenge when directly measuring location accuracy is to obtain a suitable value for ${}^{p}P^{*}_{0}$, the true position of the landmark. With the three- and seven-pixel-diameter circular landmarks tested, the CRLB shows a minimum uncertainty of about 20 millipixels, so ${}^{p}P^{*}_{o}$ must be determined to within one or a few millipixels for the accurate measurement of ${}^{p}\widetilde{P}_{o}$. This is achieved by surrounding the primary landmark with a secondary landmark that is much larger, and thereby presents many more pixels for location estimation. Furthermore, the secondary landmark is designed so that some possible bias effects are cancelled by symmetry, and also so that an internal measure of accuracy is provided by redundancy.

Chapter 7 is organized as follows. Section 7.1 presents the details of the experimental design and setup. Section 7.2 introduces the innovative experimental methodology that allows the determination of landmark location ${}^{p}P^{*}_{o}$ with very high accuracy. The camera calibration and focusing procedure is presented in Section 7.3, followed by imager noise characterization in Section 7.4. Finally, the experimental tool that enables experimental validation is presented in Section 7.5, followed by experimental results in Section 7.6.

7.1 Experiment Design and Setup

Experiments have been performed on an isolated optical bench using a Pulnix TM-9701 monochrome digital camera and 18- and 35-mm double Gauss lenses. The camera characteristics are listed in Table 7.1. The data for each experiment comprised 1800 images, three images at each of 600 positions. The landmark

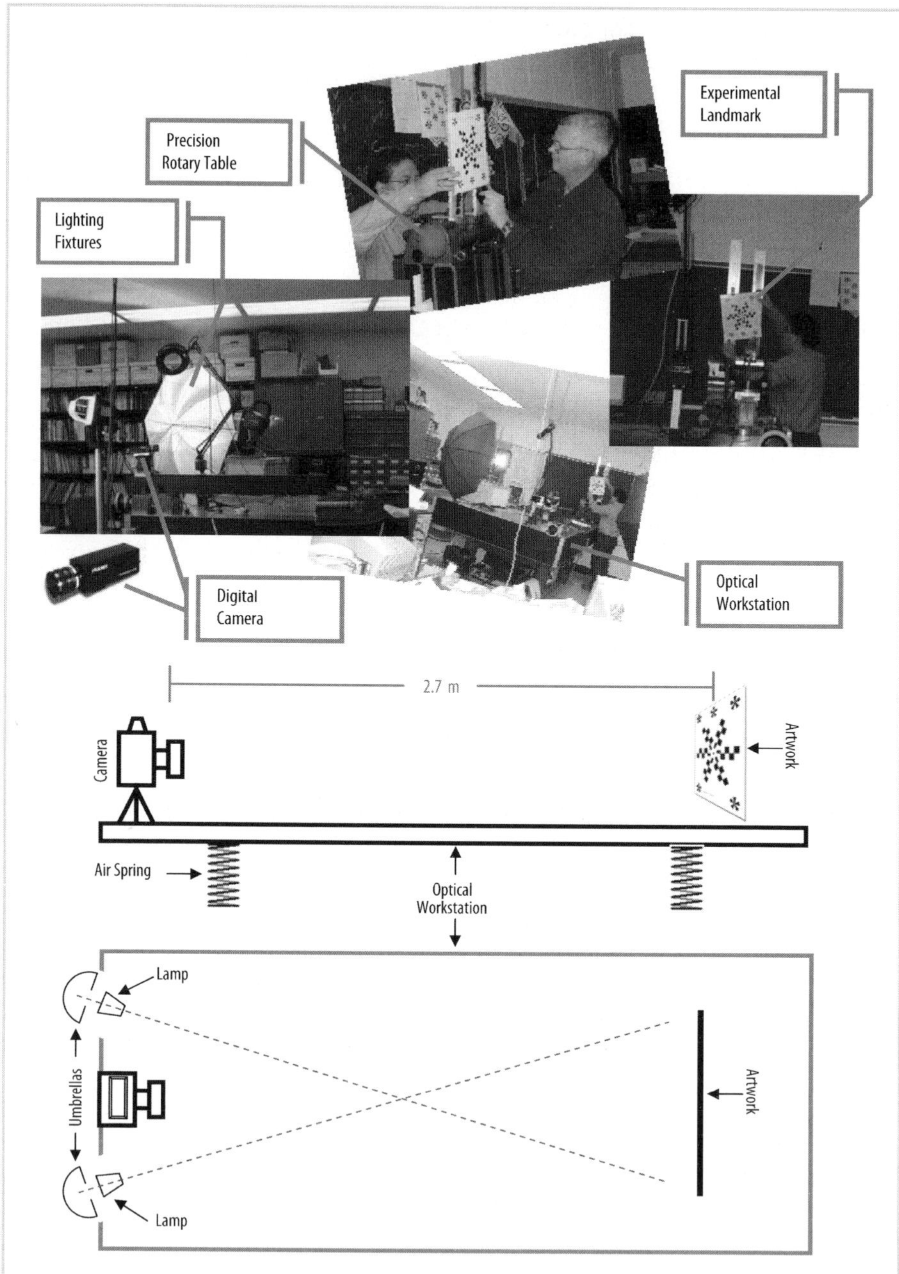

Figure 7.1. Experimental setup

Table 7.1. Pulnix TM7901: imager specifications

Imager Type	Monochrome 2/3" Progressive Scanning Interline Transfer CCD
Resolution	768 (H) × 484 (V)
Amplitude Quantization	8 bits
Pixel Size	13.6 µm × 11.6 µm
Sensitive Area	Imager Native 25% Enhanced: 65% (using microlenses)

artwork was produced on acetate film with a high-resolution laser image setter and bonded onto a glass substrate for rigidity. This was mounted to a precision two-axis rotary table. The 600 positions were randomly selected over a range corresponding to ± 1.5 pixels to achieve Monte Carlo sampling.

The camera was installed on a leveled optical bench that was supported using damped pneumatic isolators (air springs) to minimize the transmission of vibrations from the floor. The landmark was installed 2.7 meters from the principal point of the camera, on a precision two-axis rotary table installed on the optical workstation; the size of the landmark was approximately three pixels wide with the 18-mm lenses and seven pixels wide with the 35-mm lenses. To minimize the degree of nonuniform illumination on the artwork, two light sources were located symmetrically immediately behind the camera plane. Using photographer's umbrellas, the beams were directed toward the target. Details of the experimental setup are shown in Figure 7.1.

7.2 Experimental Artwork and Measuring Landmark Location Accuracy

For these experiments the primary landmark is a 6-mm-diameter circle. This landmark was embedded in a much larger secondary landmark pattern, which is shown in Figure 7.2. To create the precision artwork for this study, a commercial high-resolution printing process (called image setting) was used. This system uses a scanning laser to draw eight-micron dots in a photosensitive emulsion on an acetate film. In other applications, the image setter artwork has been shown to maintain dimensional precision to within ten microns over distances as great as 50 centimeters. The printed acetate film was then attached to a clear flat glass using a UV-cure optical adhesive. In order to avoid shadows, the printing mode was "left-reading;" that is, the photoemulsion faces the glass.

The secondary landmark artwork was designed to enable the use of a specially created estimator that provides a high-precision location reference. This estimator, called "nano-triangles," calculates the intersections of the lines formed by the three patterns surrounding the primary landmark. Ideally the three lines will intersect at a point on the imager. In practice, uncertainty in estimating the locations of the lines produces a triangle, as shown in Figure 7.3. The size of the triangle provides a measure of the uncertainty of the nano-triangles landmark location.

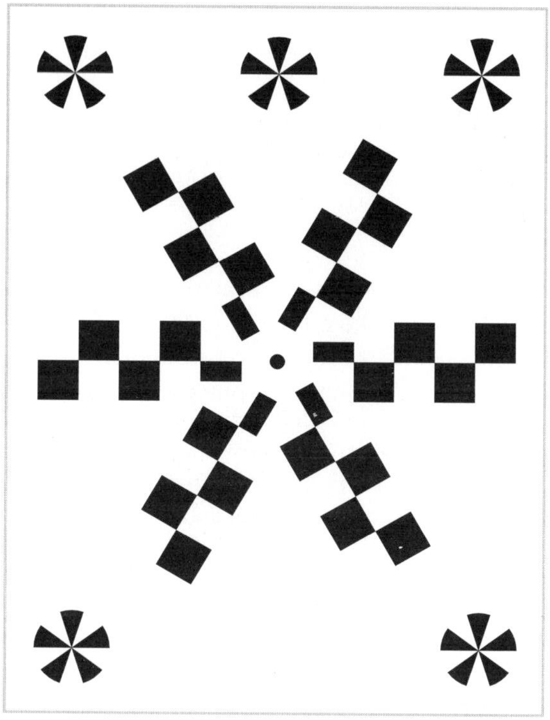

Figure 7.2. Landmark location experiment artwork design

The artwork is composed of three line-location patterns. Each of these patterns is composed of pattern boxes whose central edges are collinear with the pattern line used for high-precision location. These definitions are shown in Figure 7.3.

In order to locate the three pattern lines, the central edge in each of the line-location patterns must be found. There are a variety of edge detection algorithms (Heath *et al.* 1996), but to ensure the lowest estimation uncertainty, a methodology generalizing the Nalwa–Binford subpixel edge location algorithm was developed (Nalwa and Binford 1986). The features added to the Nalwa–Binford algorithm are:

- Detection of multiple edges around the same line.
- Use of selected pixels for edge location estimation (the original Nalwa–Binford algorithm operates with data from a 5×5 pixel image region incorporating the edge).
- New multistage seeding of the nonlinear line location estimator, as described below.

The generalized Nalwa–Binford algorithm models the central edges of the high-precision line location patterns with a hyperbolic tangent of the form:

$$L(z) = K_{nb} \pm S_{nb} \tanh\left(f_{nb}(z-p)\right); \quad z = x\sin\alpha + y\cos\alpha\,, \tag{7.2}$$

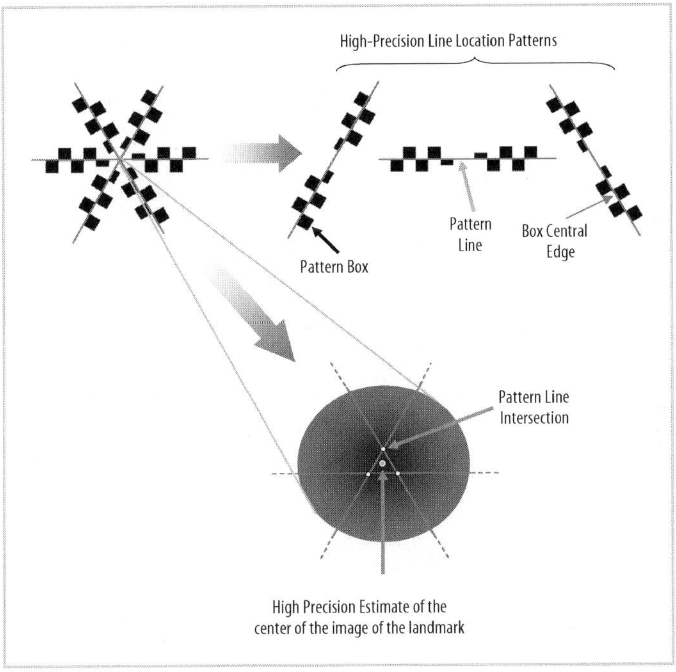

Figure 7.3. Definitions of the artwork features used in the discussion of the "nano-triangles" estimator

where K_{nb} is the mid-level of the edge with units of intensity, S_{nb} is the intensity difference between the light and dark sides of the line, f_{nb} is a constant determined by the smoothing of the imaging system (in mm^{-1}), *i.e.*, σ_b, p is the distance of the edge with respect to the origin of the coordinates (in mm) and finally α is the direction of intensity variation on the edge (in degrees). These definitions are illustrated in Figure 7.4.

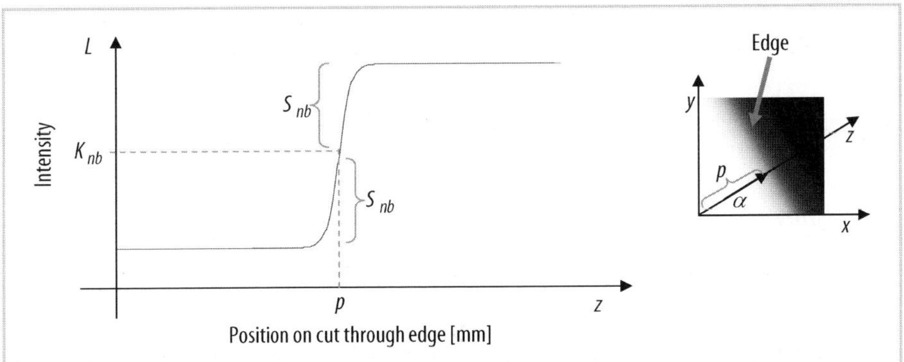

Figure 7.4. Hyperbolic tangent model for edges

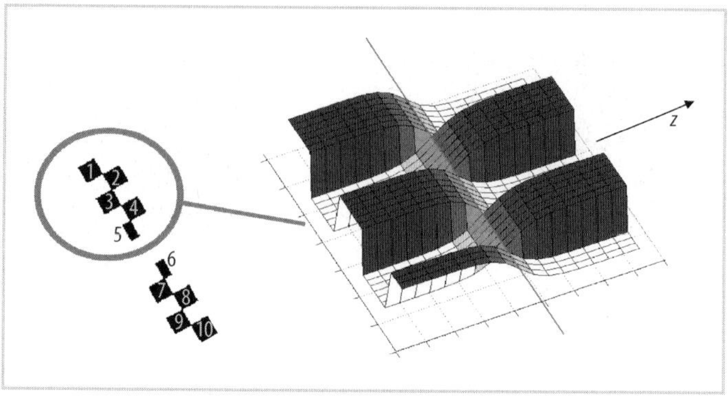

Figure 7.5. Surface pictorial of a portion of the artwork pattern. Illuminance intensity is indicated by the z coordinate of the mesh plot. As with the primary landmark, the illuminance makes a smooth transition from the high to low level.

An alternating box pattern is used in the artwork to compensate for any first-order bias in the edge location estimator. A mesh plot of the artwork image is shown in Figure 7.5, where the height at each point indicates the illuminance function at the corresponding point in the image. The symmetric alternation of pattern boxes between the left and right sides of the line tends to cancel any bias present in the edge location estimate.

The model used for the estimation of each of the three pattern line locations is:

$$L(z) = \sum_{j=1}^{5} \left[K^j_{nb} + S^j_{nb} \tanh\left(f_{nb}(z-p)\right)\right]\Big|_{\text{box}=2*j-1} + \left[K^j_{nb} - S^j_{nb} \tanh\left(f_{nb}(z-p)\right)\right]\Big|_{\text{box}=2*j}$$

$$z = x \sin\alpha + y \cos\alpha \,. \tag{7.3}$$

In this model there are several parameters K^j_{nb} and S^j_{nb} corresponding to the illumination intensity levels in neighboring pattern boxes. These parameters account for the effects of nonuniform illumination across the entire length of the pattern line. With the model presented in Equation 7.3, a nonlinear least squares estimator with 12 parameters, $\theta = K^1_{nb}, S^1_{nb}, K^2_{nb}, S^2_{nb}, K^3_{nb}, S^3_{nb}, K^4_{nb}, S^4_{nb}, K^5_{nb}, S^5_{nb}, p, \alpha$, is solved using the Newton–Raphson method.

The three pattern lines give three intersection points, as illustrated in Figure 7.6. The estimated location of the center of the landmark is calculated as the centroid of the intersection points. The uncertainty in this estimation is calculated as the norm of the vector formed between the estimated center and the farthest intersection point from it. Examples of the output from the nano-triangles estimator for two experimental images are shown in Figure 7.6.

The three edges that form the triangles seen in Figure 7.6 are independently estimated from the pattern lines of the secondary landmark, and would ideally intersect at a point. The size of the triangle is an indication of the uncertainty

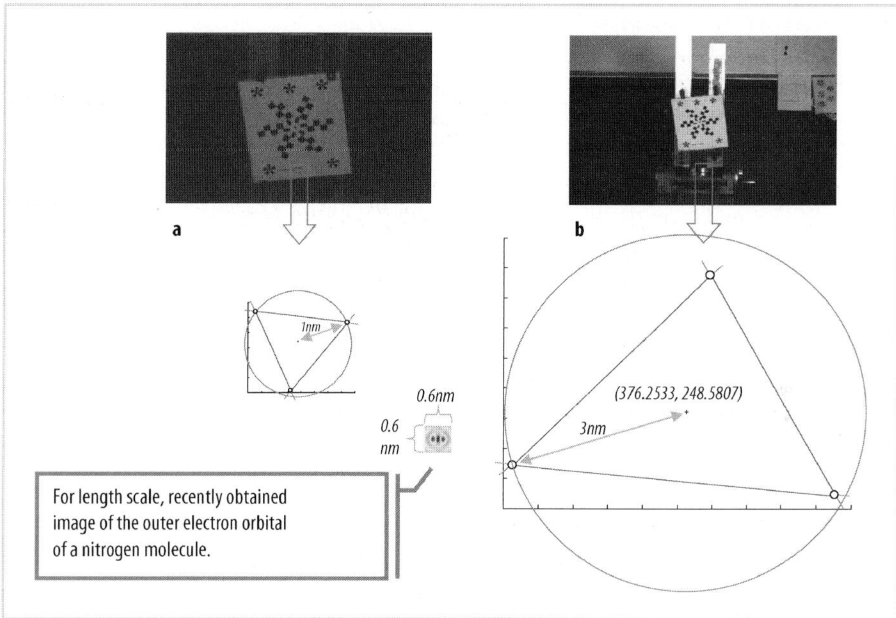

Figure 7.6a,b. Example output of the nano-triangles estimator: **a** 35-mm image; **b** 18-mm image. The one and three nanometer indications are the typical dimensions of the triangles formed, as measured on the surface of the imager. One nanometer is approximately 100 micropixels.

in the location of the image of the center of the secondary landmark. As seen in Figure 7.6, the misfit triangles have dimensions of approximately one nanometer on the imager (or approximately 0.1 millipixels) when imaged with the 35-mm lens, and three nanometers when imaged with the 18-mm lens. To highlight the accuracy of the nano-triangles location estimate, Figure 7.6 also shows a picture of the outer electron orbital of a nitrogen molecule taken recently at the Steacie Institute for Molecular Sciences[1]. As a separate measure of the secondary landmark location uncertainty, 30 images were taken with each lens at a single pose, and the secondary landmark was located in each image. These sets show a one-sigma variability of two millipixels for the 35-mm lens configuration and five millipixels for the 18-mm lens configuration. While significantly different from each other, both the triangle dimensions and the image-to-image variability support the conclusion that the reference locations given by the nano-triangles estimator are sufficiently accurate to be used to experimentally measure the primary landmark location uncertainty.

[1] Nitrogen molecule picture courtesy of Dr. David Villeneuve from the Steacie Institute for Molecular Sciences, part of the National Research Council in Canada. Reproduced with permission.

7.3 Camera Calibration

The Pulnix camera used in these experiments was calibrated using the Photomodeler Pro Version 4 camera calibration tool. This calibration was performed for each of the lenses used (18 mm and 35 mm). The outputs of the calibration procedure are the following: lens distortion parameters K_1, K_2, P_1, P_2, the adjusted focal length c_p, the coordinates of the principal point (x_o, y_o), and the imager size. An example of a display from the camera calibration tool is shown in Figure 7.7.

The camera calibration procedure involved taking pictures of a specially designed pattern in several camera poses and then performing semi-automated marking of points in all of the images. After this stage, the calibration tool applies photogrammetric bundle adjustment to the marked images to determine, among other parameters, the lens calibration. An example of the 35-mm image set used for camera calibration is shown in Figure 7.8.

Prior to calibration, the lens focus was set to a distance of 2.7 meters. This task was performed by adjusting the lens focus to a point where the modulation transfer function (MTF) of the camera plus lens system was maximized. Figure 7.9 shows the spatial pattern used to measure the image MTF. The focus control was locked and left unchanged for the duration of the experimental validation.

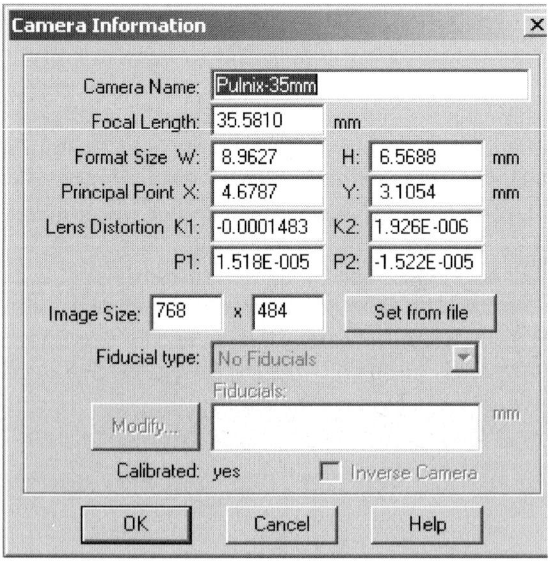

Figure 7.7. Example display of the camera calibration tool

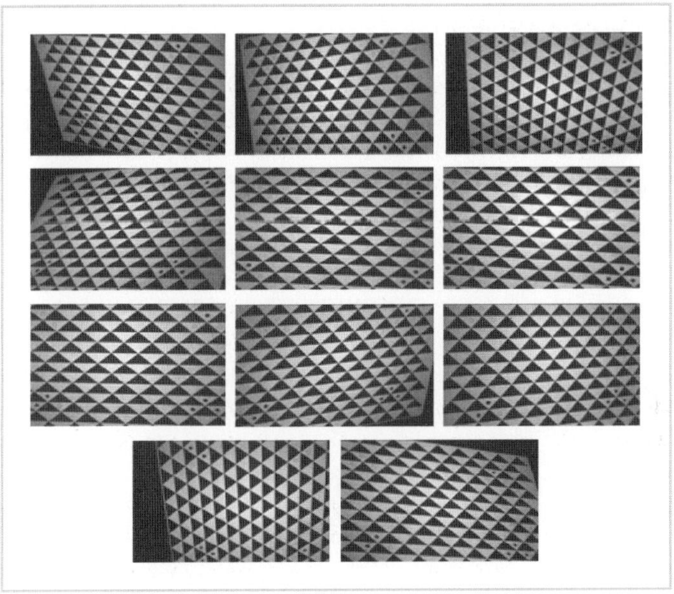

Figure 7.8. Camera calibration images for 35-mm lens

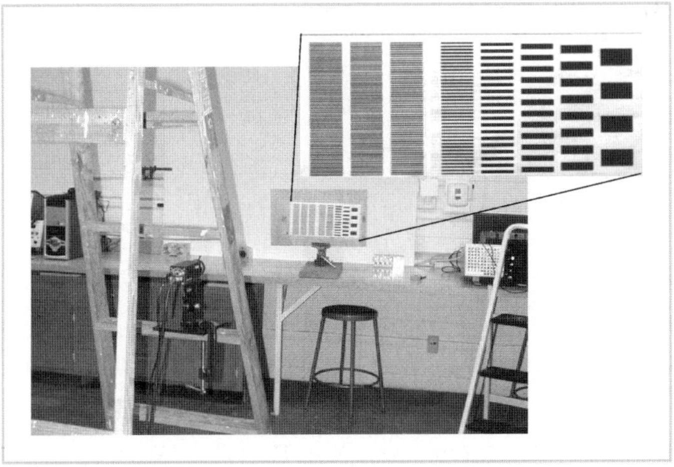

Figure 7.9. Picture of spatial pattern used for camera focusing

7.4 Imager Noise Characterization

A study of the imager noise behavior of the Pulnix TM9701 is presented in this section. The characterization is performed by calculating the RMS standard deviation $\widehat{\sigma_{\bar{J}}}$ of the pixel values from a given image captured M times. Assuming an imager with a (N_x, N_y) pixel resolution, the RMS variance is given by

$$\widehat{\sigma_{\bar{J}}^2} = \frac{\sum_{j=1}^{N_x} \sum_{k}^{N_y} \sum_{i=1}^{M} \left(\bar{I}_{(j,k,i)} - E\left\{\bar{I}_{(j,k)}\right\}\right)^2}{N_x N_y (M-1)} , \tag{7.4}$$

where $\bar{I}_{(j,k,i)}$ is the ith measured intensity of the (j,k) pixel and $E\left\{\bar{I}_{(j,k)}\right\}$ is the expected value of the measurements,

$$E\left\{\bar{I}_{(j,k)}\right\} = \frac{\sum_{i=1}^{M} \bar{I}_{(j,k,i)}}{M} . \tag{7.5}$$

Application of Equation 7.4 to a set of images acquired using the Pulnix TM9701 with a 35-mm lens is shown in Figure 7.10. It is important to highlight the fact that

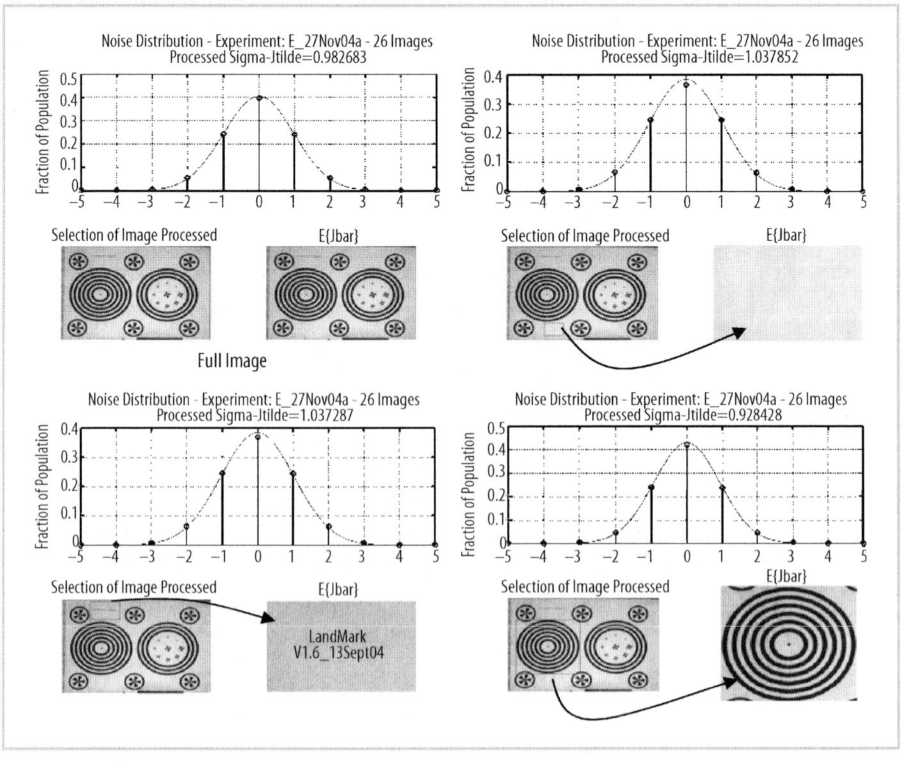

Figure 7.10. Pulnix TM9701 noise characterization: 35-mm lens

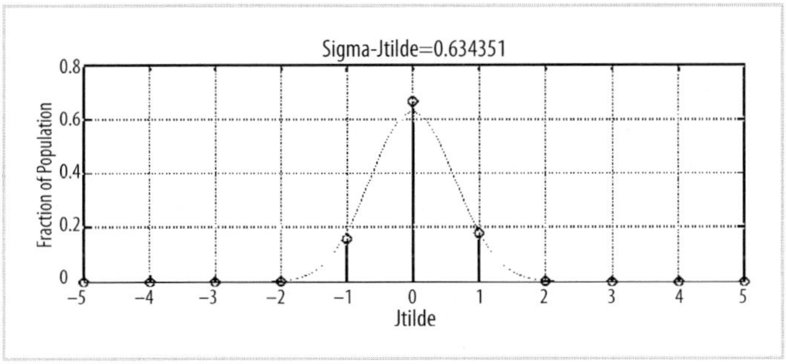

Figure 7.11. Pulnix TM9701 noise characterization: dark image

the study was performed for the full image, as well as for sections of it, in order to investigate the possibility of nonuniform noise behavior in the imager.

This characterization confirms the hypothesis of Gaussian-distributed noise and reveals that $\widehat{\sigma_{\tilde{J}}}$ is approximately one count of the Pulnix TM9701 eight-bit imager. An equivalent outcome was obtained when the same experiment was performed using an 18-mm lens.

Figure 7.11 shows the measurement of $\widehat{\sigma_{\tilde{J}}}$ using dark images (*i.e.*, with a lens cover installed). Notice that the dark current noise level is about 40% lower than that of the illuminated images. This indicates a coupling between the noise level and the illumination intensity captured on the imager, which is expected to have a contribution due to shot noise. For this research, a simplified noise model which neglects the intensity-dependent component of $\widehat{\sigma_{\tilde{J}}}$ is used.

7.5 Experimental Tool

A computational tool for automatically computing estimation confidence intervals was developed in order to realize the experimental validation. This tool, called EyeExpBox, was built using MATLAB®. EyeExpBox provides a framework for calculating landmark location uncertainty in experimental images. To accomplish this, EyeExpBox processing is divided into a set of computational components implemented in MATLAB® program files. The high-level architecture of EyeExpBox is shown in Plate 8 (see Color Section, page 151).

The first component of EyeExpBox is ProcLA. This photogrammetric tool determines the pose of the artwork with respect to the camera, enabling artwork features to be accurately located during further processing. In addition, this tool creates and stores other data structures related to the experimental configuration and camera information, including camera calibration parameters.

Image files processed by ProcLA are then passed to EyeExpConfig. This tool has several functions. First, it creates the file structure for each subsequent step of EyeExpBox. Second, it adds ancillary configuration information such as camera model, landmark type, and experimental description, to the data structures created by ProcLA. Finally, EyeExpConfig computes a high-precision estimate of the center of the landmark image by performing the nano-triangles estimation, as explained in Section 7.2. The results of this estimation are then stored for further processing in the subsequent set of EyeExpBox tools.

Figure 7.12 shows the display created by EyeExpConfig while processing one of the experimental images. Notice the marking of the corners of the pattern boxes. These points are located using the target-pose estimate provided by ProcLA. The circular areas around the boxes' center edges indicate the pixels that are used by the nano-triangles estimator.

EyeExpDisk is the tool responsible for separating the image of the landmark from the image of the artwork. This is done by estimating the landmark center using the nano-triangles algorithm, and then extracting the set of pixels within a radius from this point. The radius is calculated as the mid-point between the outer edge of the landmark and the closest pattern box. Figure 7.13 shows an example of the landmark separation process executed by EyeExpDisk.

The next tool within EyeExpBox is EyeExpT. This tool applies landmark location estimators to the separated images. As in the case of EyeT (see Section 6.4), the estimators used in EyeExpT are binary centroid, grayscale centroid, ellipsoidal con-

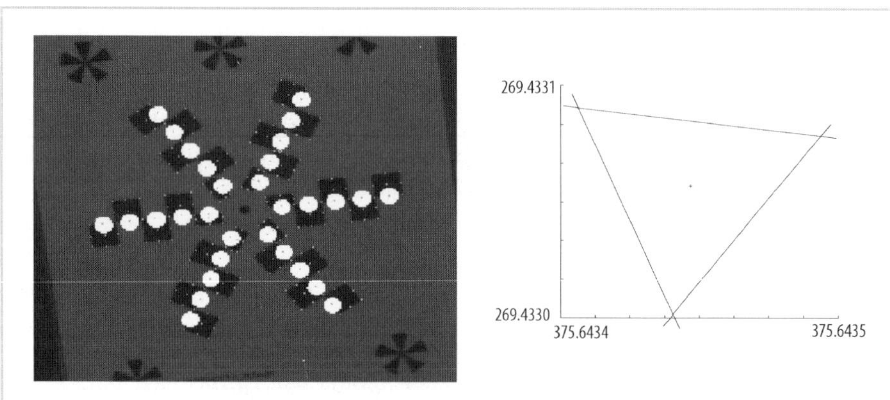

Figure 7.12. Display generated by EyeExpBox. *Bright circles* show pixels selected for processing by the nano-triangles estimator.

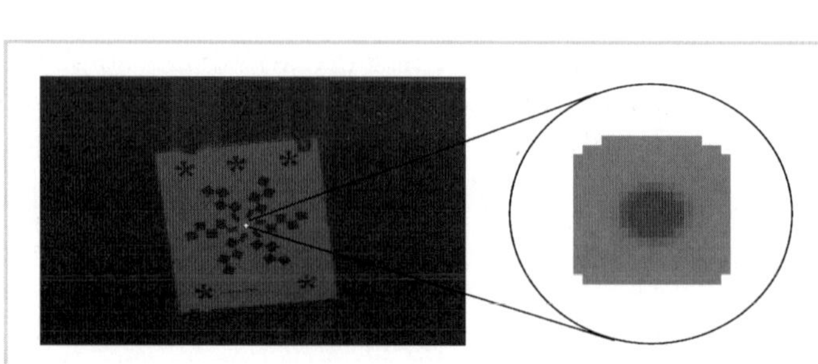

Figure 7.13. Landmark separation performed by EyeExpDisk

tour, and Butterworth tepuy. In line with photogrammetric practice, the landmark location estimators are implemented in image coordinates.

Finally, EyeExpCov processes the estimates from the EyeExpT and generates experimental estimation confidence intervals. The confidence interval for a given estimator is calculated as follows:

- For each estimate from the Monte Carlo sampling, form:

$$\widetilde{P}_{\underset{L}{o}}(j,T) = \widehat{P}_{\underset{L}{o}}(j,T) - \widehat{P}^{*}_{\underset{L}{o}}(j) \quad 1 \le j \le M\,, \tag{7.6}$$

  where $\widehat{P}^{*}_{\underset{L}{o}}(j)$ is the jth estimate of the secondary landmark location obtained from the nano-triangles estimator from the M experimental repetitions, $\widehat{P}_{\underset{L}{o}}(j,T)$ is the estimated center of the landmark image from the jth image, obtained using estimator T, and $\widetilde{P}_{\underset{L}{o}}(j,T)$ is the misadjustment of the estimator.
- The calculated misadjustments are sorted into order of ascending magnitude, $\left\|\widetilde{P}_{\underset{L}{o}}(k-1,T)\right\| \le \left\|\widetilde{P}_{\underset{L}{o}}(k,T)\right\| \le \left\|\widetilde{P}_{\underset{L}{o}}(k+1,T)\right\|$ with $1 < k < M$.
- The confidence interval of estimator T is given as:

$$\overline{C_n}^{T} = \left\|P_{\underset{L}{o}}(k,T)\right\|\,, \quad k = n * M \qquad 0 < n \le 1\,. \tag{7.7}$$

Example displays from the EyeExpCov tool are presented in the next section.

7.6 Experimental Results

This section describes experimental results for the confidence interval and the corresponding probability density function of the studied estimators. The study was conducted using the two configurations indicated in Section 7.1, processing 1800 images; 600 poses and three images acquired at each pose.

18-mm Configuration

The confidence interval results from the 18-mm configuration are shown in Plate 9 (see Color Section, page 152). These confidence intervals are computed as the rings centered around the estimated true center of the landmark image that encloses 95% of the Monte Carlo samples for each estimator used.

Plate 10 (see Color Section, page 152) displays the same confidence intervals without the Monte Carlo sample details, highlighting $E\left\{{}^{P}\widehat{P}_{o}\right\}$ for each of the estimators used.

Figure 7.14 shows the experimentally determined distributions of landmark location error obtained using an 18-mm lens.

As expected, the binary centroid estimator presents the greatest uncertainty. This is followed by the grayscale centroid estimator, which substantially reduces

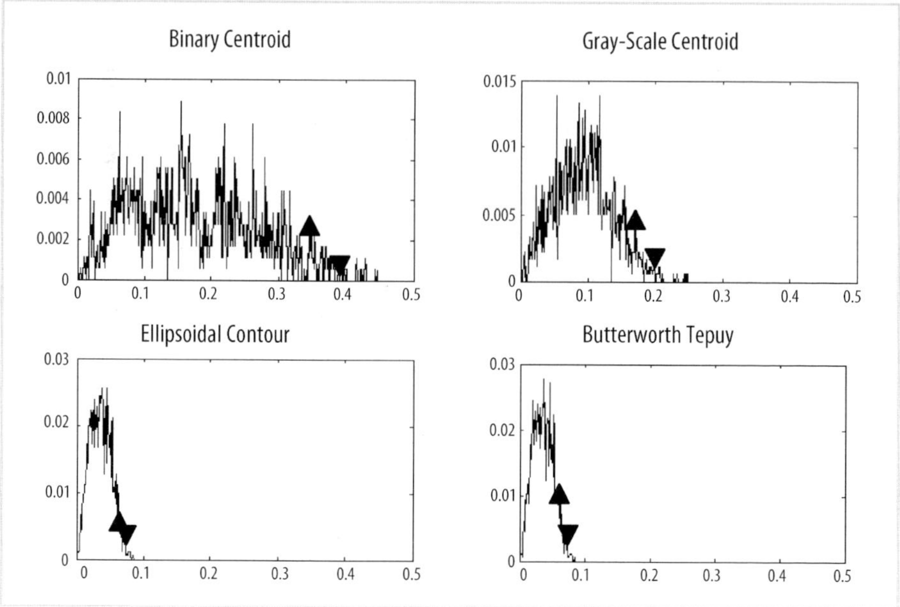

Figure 7.14. 18-mm configuration: experimentally determined distributions of landmark location error

uncertainty. The 95% confidence level for the binary centroid algorithm is 345 millipixels, while it is 169 millipixels for the grayscaled centroid.

The model-based ellipsoidal contour and Butterworth tepuy estimators show much lower uncertainties than the centroid estimators. The model-based estimators show near-identical performance, with a 95% confidence interval of 63 millipixels. The reduction in estimator bias achieved by explicitly modeling the landmark shape is a chief contributor to the improvement. The Butterworth tepuy algorithm shows a performance that is few millipixels better than the ellipsoidal contour estimator, even though the computational cost of the Butterworth tepuy algorithm is approximately 30 times greater than that of the ellipsoidal contour.

35-mm Configuration

The confidence interval results for the 35-mm configuration are shown in Plate 11 (see Color Section, page 153). This plate also displays $E\left\{{}^{P}\widehat{P}_{o}\right\}$ for each of the estimators used. Figure 7.15 shows the experimentally determined distributions of landmark location error obtained using the 35-mm lens.

These results reveal a similar behavior to the one obtained for the 18-mm configuration experiment. The binary centroid estimator displays a 95% confidence level near 300 millipixels, while the grayscale centroid estimator features the same confidence level at 260 millipixels.

The ellipsoidal contour and Butterworth tepuy estimators continue to show a much higher performance, with 95% confidence levels of 52 and 50 millipixels,

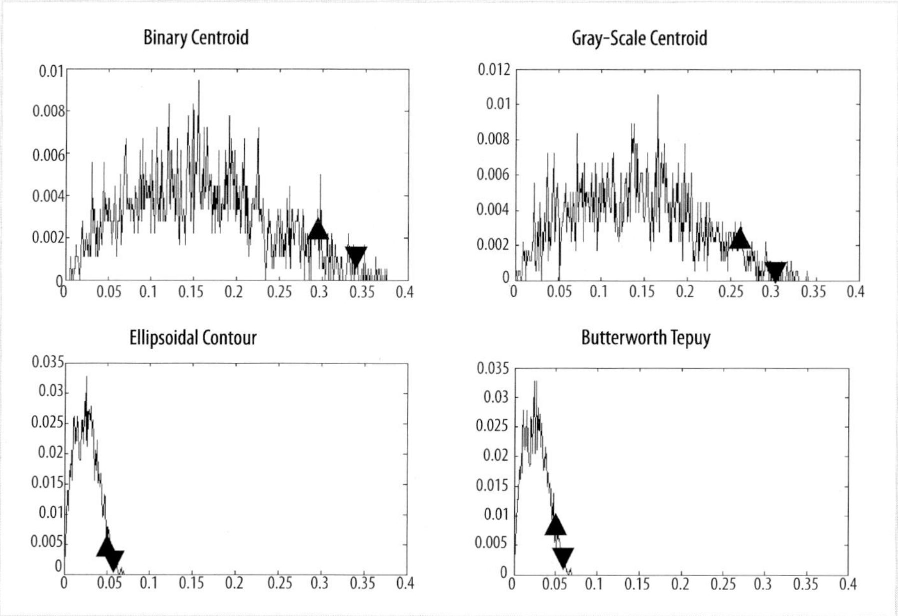

Figure 7.15. 35-mm configuration: experimentally determined distributions of landmark location error

respectively. Again, the performance of the Butterworth tepuy estimator is slightly better than that of the ellipsoidal contour estimator.

Plates 10 and 11 show the existence of a bias of approximately 0.015 pixels for the higher-accuracy ellipsoidal contour and Butterworth tepuy estimators (and larger for the centroid estimators). Several hypotheses were considered to explain the bias, including an offset in the artwork, nonuniform lighting, and an imager-related artifact. Experiments were designed to test each hypothesis and data were collected:

- Rotating the artwork by 180° while keeping the rest of the configuration intact
- Rotating the camera by 180° while keeping the rest of the original configuration unchanged
- Modifying the lighting

However, the results obtained were inconclusive. It is believed that the residual biases are due to a combination of effects.

The nano-triangles artwork and estimation algorithm gives a reference in the image with an uncertainty of two millipixels in the 35-mm configuration and five millipixels in the 18-mm configuration, which is much less than the uncertainty primary landmark location estimator in each case. This experiment provides the means to directly measure the uncertainty of the primary landmark location estimators. Measurements show that the model-based estimators perform with significantly lower uncertainty than the centroid estimators, providing a five- to sixfold improvement.

8 Studies of Landmark Location Uncertainty

The EyeBox tool provides the means to examine the potential and actual landmark location uncertainty as a function of landmark and imager configuration. In this chapter, the tool is employed to investigate the interactions of imager noise, smoothing radius, luminosity dynamic range, landmark size and tilt, imager pixel sensitive area and aspect ratio, nonuniform illumination and imager amplitude quantization on landmark location uncertainty. These results are presented in the form of performance maps showing the size of the 95% uncertainty confidence

Table 8.1. Baseline configuration for the experimental and numerical investigations

	Parameter	Value	Description
Camera	Camera Name	PULNiX9701	Imager mode based on the Pulnix 9701
	Lens	18mm/35mmDG	Double Gauss lens
	Amplitude Quantization	8 bits	Quantization level over dynamic range
	N_x	768	Imager resolution: horizontal [pixels]
	N_y	484	Imager resolution: vertical [pixels]
	Sblur	0.009	Diffraction blur radius [mm]
	Std. Deviation Noise	0.004	Standard deviation of noise in imager: $1/2^8$
	w_x	80%	Percentage pixel sensitive area on x axis
	w_y	80%	Percentage pixel sensitive area on y axis
	k_x	83	Pixel density in x. $k_x = N_x$/width [pixel/mm]
	k_y	73	Pixel density in y. $k_y = N_y$/height [pixel/mm]
	c_p	18/35	Principal distance [mm]
Landmark	cR_L	3	Landmark radius [mm]
	cX_L	0	x coordinate of landmark center in camera coordinates [mm]
	cY_L	0	y coordinate of landmark center in camera coordinates [mm]
	cZ_L	2700	z coordinate of landmark center in camera coordinates [mm]
	cPitch_L	variable	Pitch of target in camera coordinates
	cRoll_L	variable	Roll of target in camera coordinates
	cYaw_L	variable	Yaw of target in camera coordinates
	Lmin	0.6	Luminosity dark [fraction of full scale]
	Lmax	0.9	Luminosity light [fraction of full scale]

interval, following the framework developed in Chapter 3. In each case, the experimental configuration described in Chapter 7 is taken as the baseline configuration, the parameters of which are listed in Table 8.1. In Section 8.1, analytic calculations of landmark location uncertainty are validated against the experimental measurements. In Sections 8.2 through 8.9, the analytic tools are used to investigate the impact on uncertainty of a range of configuration variables.

8.1 Theoretical and Experimental Determination of Landmark Location Uncertainty

Figure 8.1 shows the 95% confidence interval performance maps of the four estimators: the binary and grayscale centroid estimators and the ellipsoidal contour and Butterworth tepuy estimators. The performance maps include the Cramér–Rao lower bound, determined as described in Section 3.1; analytic results for each estimator, determined as described in Section 3.2; and experimental results obtained as described in Chapter 7. The data of Figure 8.1 are shown in Table 8.2.

To generate analytic results, the EyeBox configuration file was set with parameters that mimicked the experimental setup. Camera parameters were obtained from the camera calibration procedure, artwork location was obtained by the use of the nano-triangles algorithm, and landmark specification was taken from the file used

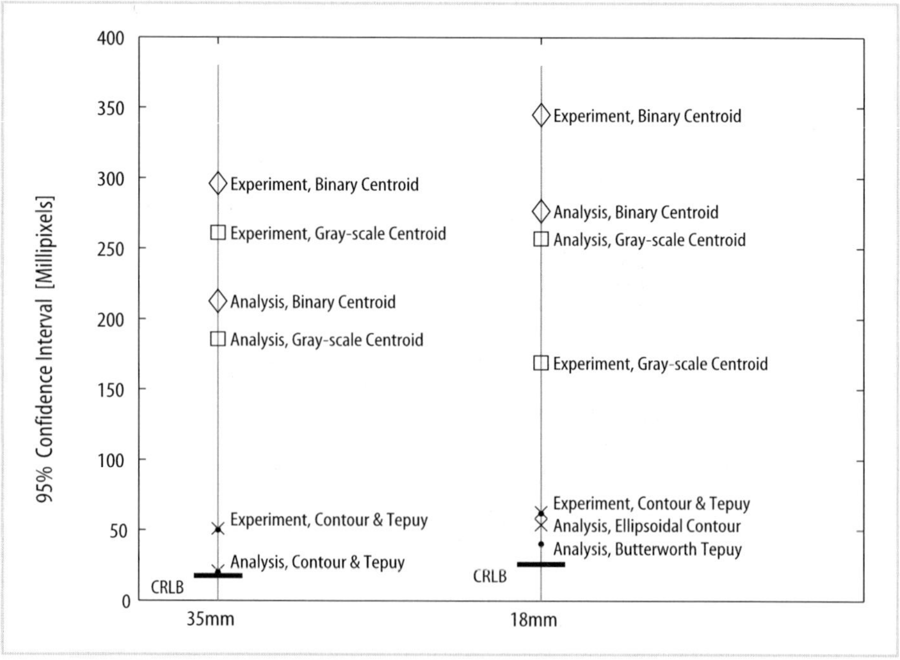

Figure 8.1. Performances of landmark location estimators; experimental and analytical determinations

Table 8.2. Radius of 95% circular confidence interval for the experimental configuration, showing values given by the analysis method of Section 3.2, the experiments of Chapter 7, and the Cramér–Rao lower bound calculation of Section 3.1; for the experimental configurations with the 35-mm and 18-mm lenses.

Method		35-mm lens	18-mm lens
Binary centroid	Analysis	212.5	276.5
	Experiment	296.0	345.0
Grayscale centroid	Analysis	185.5	257.0
	Experiment	261.0	169.0
Ellipsoidal contour	Analysis	21.0	54.0
	Experiment	51.0	63.0
Butterworth tepuy	Analysis	20.0	40.5
	Experiment	50.0	62.0
Cramér–Rao lower bound	Analysis	17.5	26.0

for the artwork generation. In addition, diffraction smoothing and illumination levels were measured using the images generated in the laboratory experiments using the Butterworth tepuy algorithm. Finally the nonuniform illumination was measured with a specially designed tool.

Figure 8.2 shows the potential and experimentally achieved circular landmark location estimation performance for a 6-mm circle imaged at 2.7 meters using 35- and 18-mm lenses in the configuration described above. At the lowest level of uncertainty is the Cramér–Rao lower bound, with an uncertainty of 20 millipixels. An idealized estimator with Cramér–Rao bound performance would give an estimate within this distance of the true image of the landmark center in 95% of trials. Analysis predicts performance levels for the model-based elliptical contour and Butterworth tepuy estimators that are indistinguishable from each other, and

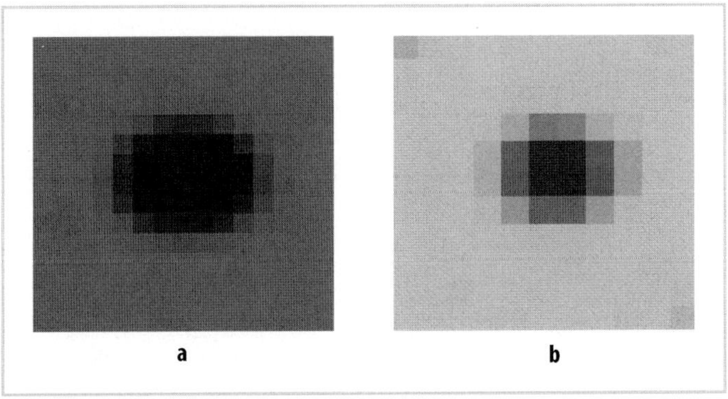

Figure 8.2a,b. Example landmark images from the experimental investigation given the results of Figure 8.1. This study focused on landmarks 4–10 pixels across. The expected precision with larger landmarks is investigated in Section 8.5. **a** Typical landmark image with 35 mm lens; **b** typical landmark image with 18-mm lens.

slightly higher than the Cramér–Rao lower bound. One of the surprises of this research is that practical estimators can so closely approach the CRLB.

The experimental estimators show a 95% confidence interval of 50 millipixels for the 35-mm lens (seven-pixel-diameter landmark) and 62 millipixels for the 18-mm lens (four-pixel-diameter landmark). These levels are $2.5\times$ and $3\times$ the analytic predictions, respectively, indicating that experimental uncertainties exist which have not yet been captured in the model. Nonetheless, the agreement between the analytic results and the experiment is striking. With the exception of the experimental result for the grayscale centroid estimator with the 18-mm lens, the experimental data follow the pattern of the analytic results in detail, with a consistently elevated level of uncertainty in the experimental result relative to analysis. Figures 8.3 and 8.4 show the joint probability distributions for each of the estimators studied for the 18-mm and 35-mm configurations. These distributions give rise to the performance maps corresponding to the analytical results shown in Figure 8.1. These distributions are the result of integration of the landmark location confidence intervals, as described in Section 3.2, and show the same structure as the experimentally generated distributions, seen in Figures 7.14 and 7.15 for the 18-mm and 35-mm lenses, respectively.

In Figures 8.3 and 8.4, it is observed that the binary and grayscale centroid distributions are not smooth. This is due to the bias characteristics of these estimators and the nonstationary statistics of landmark location estimation. Running the EyeBox tool with a more dense raster resolution would smooth these curves, but at the expense of very long computational time.

The agreement between experiment and analysis seen in Figure 8.1 and the agreement of the distributions in Figures 7.14, 7.15, 8.3, and 8.4 indicates that the analysis tools reliably capture the interactions of configuration parameters and landmark location performance. With the analytic tools validated by the agreement of the analytic results with the experimental results, it is now possible to use the EyeBox tool to investigate interactions between selected configuration parameters and landmark location uncertainty. Sections 8.2 through 8.9 present analysis-based studies of these interactions. Results from studies of the 35-mm configuration are reported. In each case similar results were obtained for the 18-mm configuration.

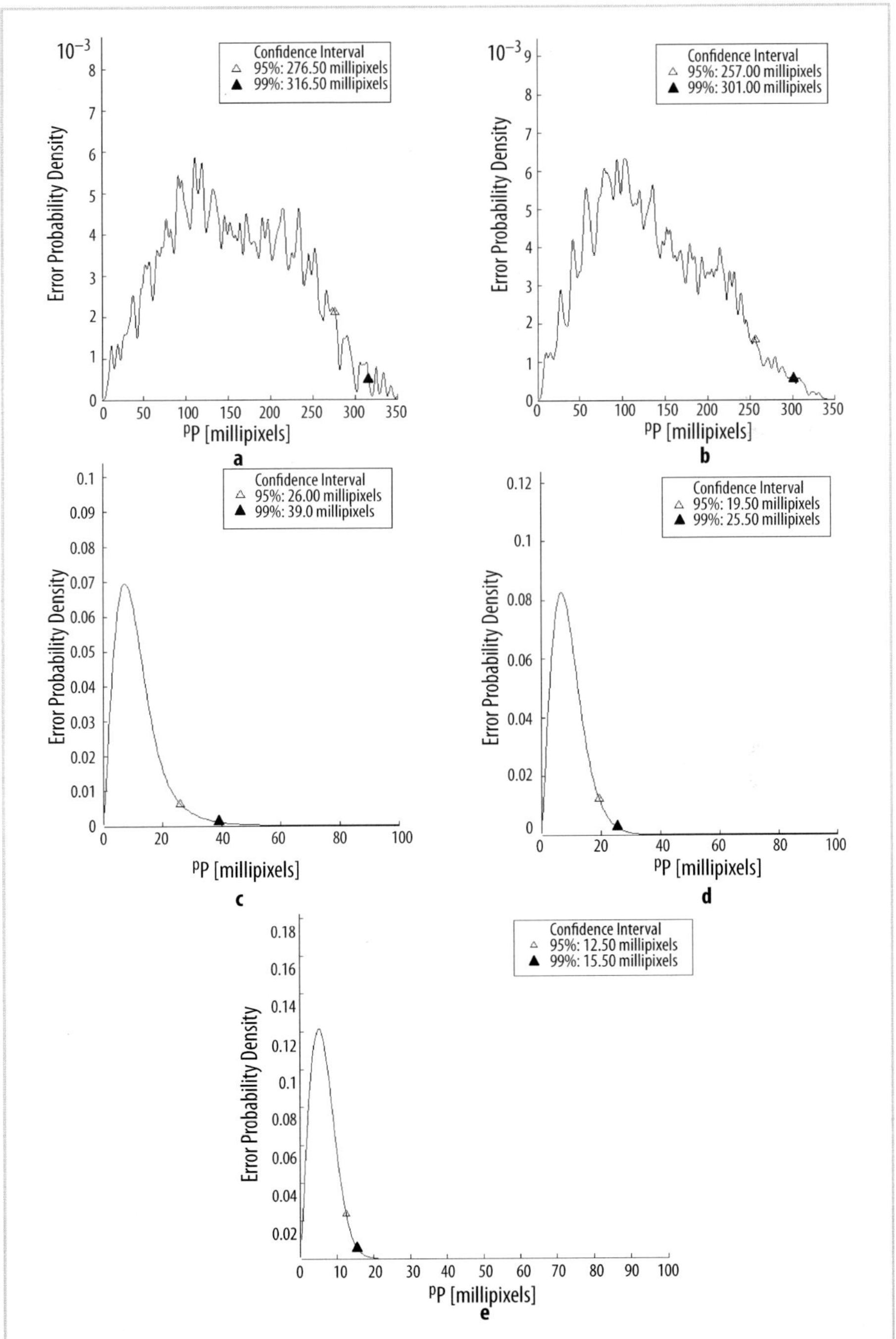

Figure 8.3a–e. 18-mm configuration: analytically determined radial probability distributions: **a** binary centroid estimator; **b** grayscale centroid estimator; **c** ellipsoidal contour estimator; **d** Butterworth tepuy estimator; **e** Cramér–Rao calculation.

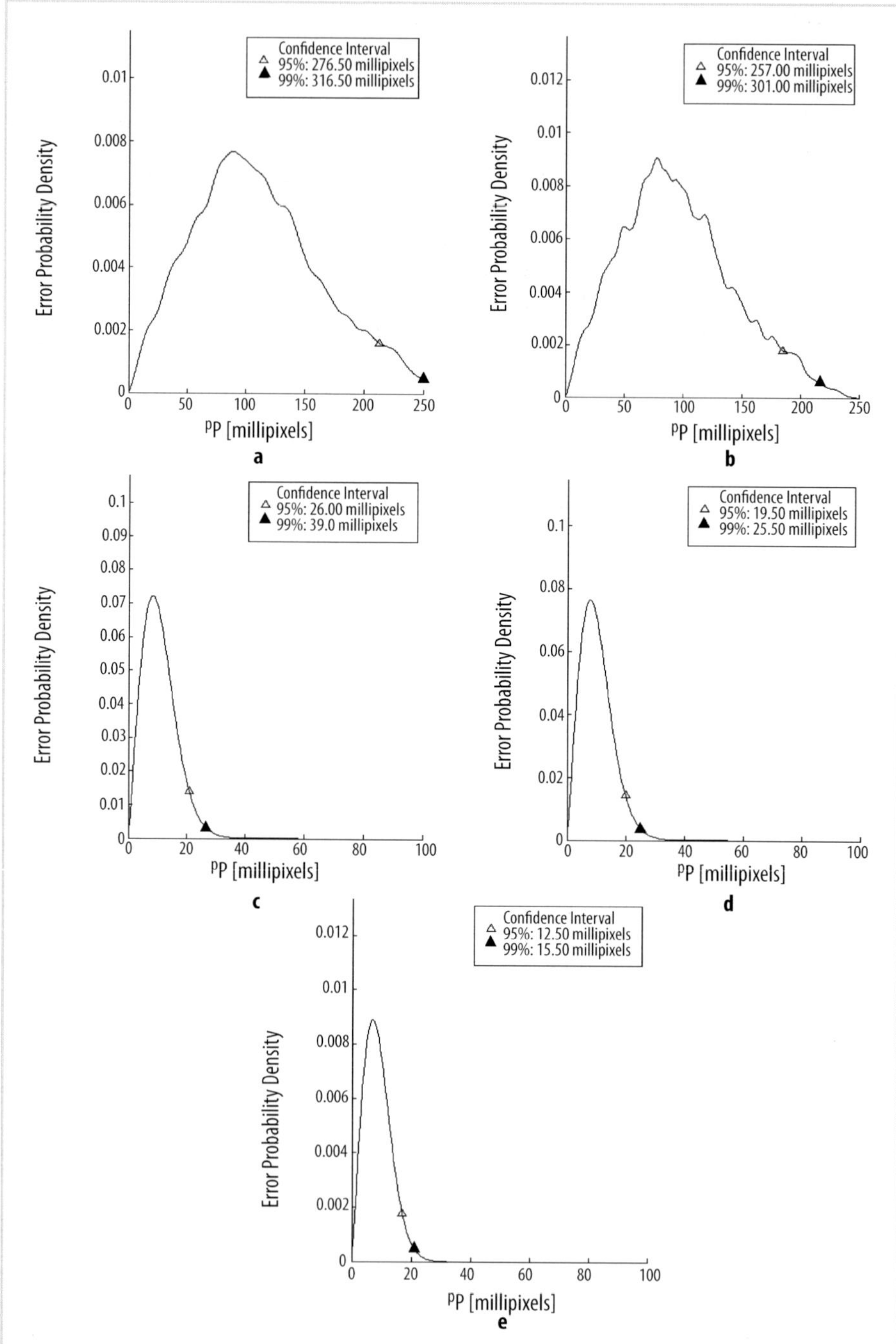

Figure 8.4a–e. 35-mm configuration: analytically determined radial probability distributions: **a** binary centroid estimator; **b** grayscale centroid estimator; **c** ellipsoidal contour estimator; **d** Butterworth tepuy estimator; **e** Cramér–Rao calculation.

8.2 Study of Effects of Imager Noise

Figure 8.5 presents the results obtained when the imager noise level was increased for the 35-mm experimental configuration. Analysis was performed for noise levels $\sigma_{\tilde{J}} = \{1, 2, 4\} \times (1/256)$ of full scale. The data for Figure 8.5 are shown in Table 8.3.

As expected, this study shows that the Cramér–Rao lower bound on uncertainty rises with as imager noise increases. The degree to which the performance of the model-based estimators tracks the Cramér–Rao lower bound is striking when it is recalled that the CRLB and practical estimator performances are obtained from very different calculations. Also striking is the degree to which the binary and grayscale centroid estimators are insensitive to noise. It is conjectured that this effect arises with the averaging that takes place in the calculation of the centroid, and the greater initial uncertainty of these estimators.

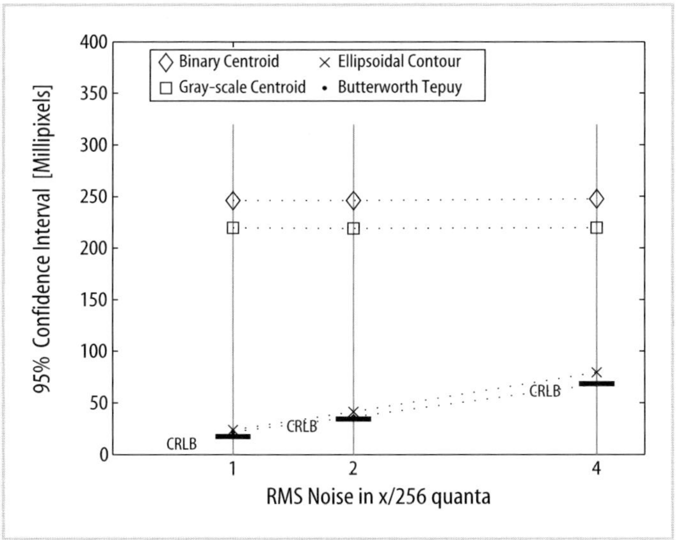

Figure 8.5. Landmark location uncertainty performance map: effects of increasing $\sigma_{\tilde{J}}$

Table 8.3. Radius of 95% circular confidence interval for three levels of imager noise, determined by analysis for four estimators and also the Cramér–Rao lower bound. Noise given in units of 1/256 full-scale, or counts of noise for an eight-bit ADC. The baseline value for the experimental configuration is 1.0 counts.

Method	Noise level (1/256 full-scale)		
	1.0	2.0	4.0
Binary centroid	246.0	246.0	247.5
Grayscale centroid	219.5	219.0	219.5
Ellipsoidal contour	24.0	41.5	79.5
Butterworth tepuy	22.0	36.5	68.0
Cramér–Rao lower bound	17.5	34.5	68.5

8.3 Study of the Effects of Smoothing Radius on Landmark Location Uncertainty

Figure 8.6 presents the results obtained when the diffraction-smoothing radius was changed for the 35-mm experimental configuration. Analysis was performed for smoothing radii of $\sigma_b = \{1, 4.5, 9, 18\}$ microns. The data for Figure 8.6 are shown in Table 8.4.

Study 8.3 shows that the smoothing introduced by diffraction lowers landmark location uncertainty (improves precision) when the smoothing radius is smaller than a pixel. The pixel dimension of the Pulnix TM9701 camera used for this study is $13.6 \times 11.6\,\mu m$. Considering the relation to pixel size, diffraction (or defocus) smoothing up to half the pixel dimension is seen to give a large improvement in precision for the model-based estimators, while smoothing up to the pixel dimension is seen to give a modest improvement for the centroid estimators.

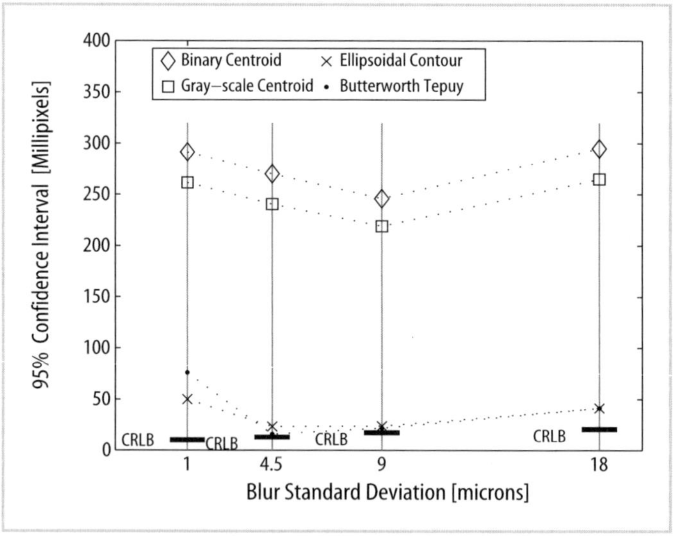

Figure 8.6. Landmark location uncertainty performance map: effects of changing σ_b

Table 8.4. Radius of 95% circular confidence interval for four values of the smoothing radius σ_b, determined by analysis for four estimators and also the Cramér–Rao lower bound. The smoothing operator radius is expressed in microns on the imager surface. The baseline value for the experimental configuration is 9.0 microns.

Method	Smoothing radius (microns)			
	1.0	4.5	9.0	18.0
Binary centroid	291.5	270.0	246.0	295.0
Grayscale centroid	261.5	240.5	219.0	265.0
Ellipsoidal contour	50.0	23.5	24.0	41.5
Butterworth tepuy	76.0	16.0	22.0	41.5
Cramér–Rao lower bound	10.0	13.0	17.5	21.0

8.4 Study of the Effects of Luminosity Dynamic Range

Figure 8.7 presents the results obtained when the luminosity dynamic range was increased for the 35-mm experimental configuration. The luminosity dynamic range is the difference between the background and the landmark luminosity, expressed as a percentage of full scale for the imager. As with the other cases, the analysis models an eight-bit imager. Analysis was performed for dynamic range values of 30%, 60%, and 80%. The data for Figure 8.7 are shown in Table 8.5.

Study 8.4 shows strongly decreasing uncertainty with increasing dynamic range. While it is expected that increasing the dynamic range will reduce uncertainty, the magnitude of this effect is surprising.

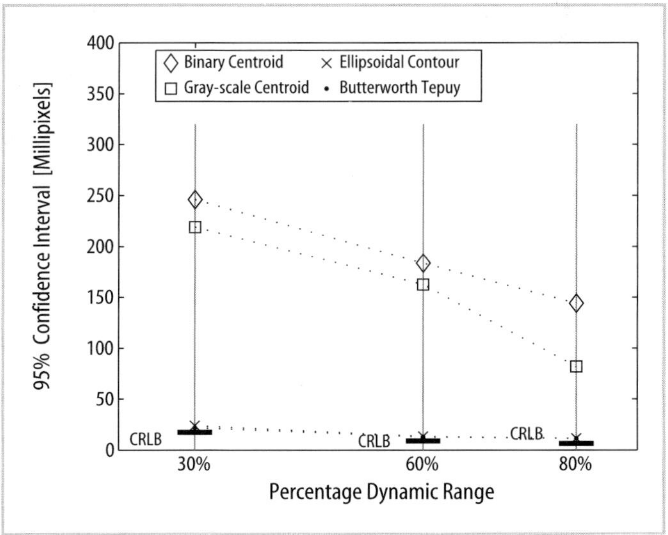

Figure 8.7. Landmark location uncertainty performance map: effects of increasing luminosity dynamic range

Table 8.5. Radius of 95% circular confidence interval, determined by analysis for four estimators and the Cramér–Rao lower bound. The luminosity dynamic range is given as the ratio of the landmark interior and exterior luminosity as percent of full scale. The baseline value for the experimental configuration is 30%.

Method	Luminosity dynamic range		
	30%	60%	80%
Binary centroid	246.0	183.5	144.0
Grayscale centroid	219.0	162.5	82.0
Ellipsoidal contour	24.0	13.5	11.5
Butterworth tepuy	22.0	13.0	11.5
Cramér–Rao lower bound	17.5	9.0	6.5

8.5 Study of the Effects of Landmark Size

Figure 8.8 presents the performance maps for different landmark sizes for the 35-mm experimental configuration. Analysis was performed for landmark radii of $\{1.5, 3.0, 6.0, 12.0\}$ millimeters. The baseline configuration (2.7 meters to target, 35-mm lens, 11.6×13.6 micron pixels) gives 1.1 mm/pixel in the horizontal direction, so the figures of $\{1.5, 3.0, 6.0, 12.0\}$ are also approximately the radii of the landmarks in pixels. The data for Figure 8.8 are shown in Table 8.6.

It is observed in Study 8.5 that increasing landmark size decreases location uncertainty for all estimators. Surprisingly, the effect seems to taper off for the centroid estimators, which reach a minimum level of uncertainty near 150 millipixels. For the landmark with a radius of 12 millimeters (a radius of approximately 12 pixels in the image), the model-based estimators achieve an expected uncertainty of 10 millipixels.

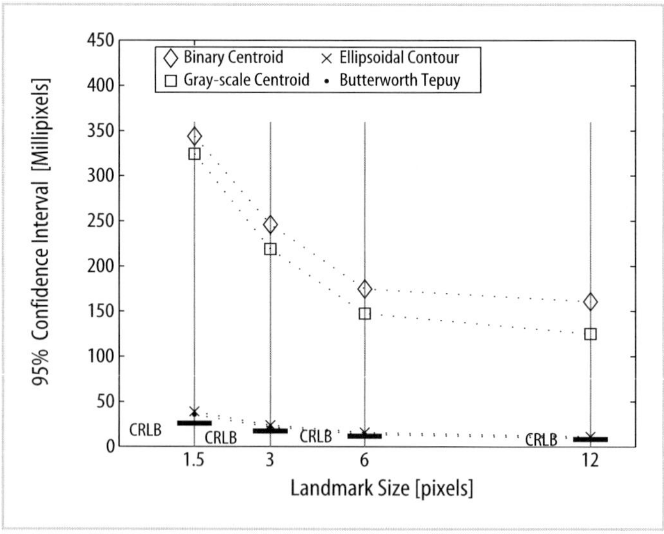

Figure 8.8. Landmark location uncertainty performance map: effects of landmark size

Table 8.6. Radius of 95% circular confidence interval for four values of landmark radius, determined by analysis for four estimators and also for the Cramér–Rao lower bound. The landmark radius is expressed in units of pixels in the image (root-mean-square radius, to consider perspective distortion or nonsquare pixels). The baseline value for the experimental configuration is 3.0 pixels.

Method	Landmark radius (RMS pixels)			
	1.5	3.0	6.0	12.0
Binary centroid	344.0	246.0	175.0	161.0
Grayscale centroid	324.5	219.0	147.5	125.0
Ellipsoidal contour	39.0	24.0	16.0	11.5
Butterworth tepuy	35.5	22.0	14.0	9.5
Cramér–Rao lower bound	26.0	17.5	12.0	8.5

8.6 Study of the Effects of Landmark Tilt

Figure 8.9 presents the results obtained when the pitch was increased for the 35-mm experimental configuration. The analysis was performed for pitch angles of 0°, 10°, 20°, and 30°.

Out-of-plane rotation, or tilt, is generally avoided in machine vision applications, but it may be unavoidable in photogrammetric or other applications that are characterized by a reduced degree of control over the presentation of the landmark for imaging. The results from Study 8.6 show that the model-based estimators are robust to landmark tilt. These estimators each include ellipsoidal eccentricity as a parameter.

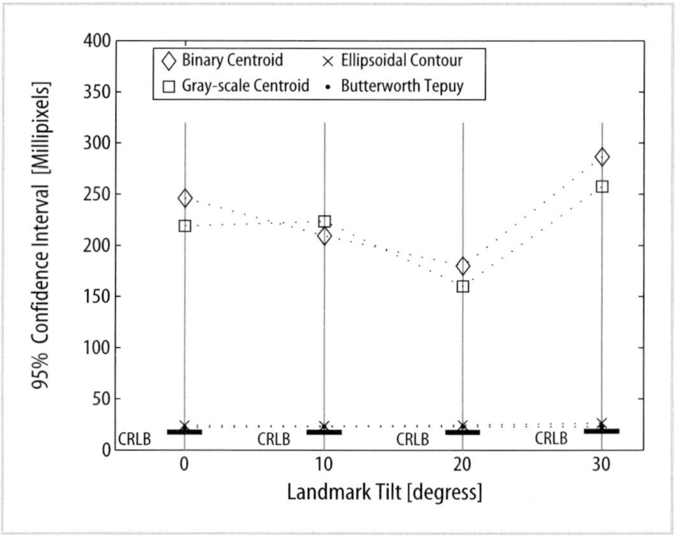

Figure 8.9. Landmark location uncertainty performance map: effects of increasing the pitch angle

Table 8.7. Radius of 95% circular confidence interval for four levels of out-of-plane rotation (tilt), determined by analysis for four estimators and the Cramér–Rao lower bound. Tilt given in degrees. The baseline value for the experimental configuration is 0 degrees.

Method	Tilt (degrees)			
	0	10	20	30
Binary centroid	246.0	209.5	180.0	286.5
Grayscale centroid	219.0	223.5	160.0	257.5
Ellipsoidal contour	24.0	23.5	24.5	26.5
Butterworth tepuy	22.0	23.0	23.0	23.0
Cramér–Rao lower bound	17.5	17.5	17.5	18.5

8.7 Study of the Effects of the Pixel Sensitive Area and Aspect Ratio

Figure 8.10 presents the results obtained when the pixel sensitive area was changed for the 35-mm experimental configuration, and Figure 8.11 presents the results obtained when the pixel sensitive area was changed for the same configuration, with the smoothing radius changed to 2.0 microns. Figure 8.12 presents the performance results for different pixel aspect ratios in the 35-mm experimental configuration. The data for Figures 8.10–8.12 are shown in Tables 8.8–8.10, respectively.

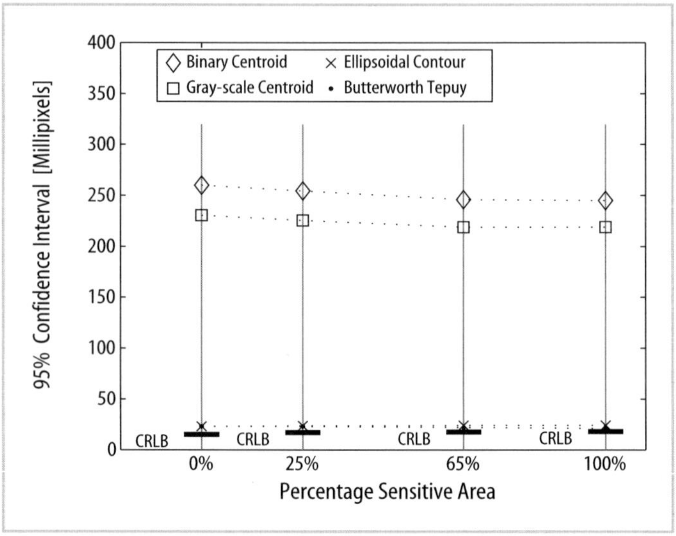

Figure 8.10. Landmark location uncertainty performance map: effects of changing the pixel sensitive area, $\sigma_b = 9.0$ microns

Table 8.8. Radius of 95% circular confidence interval for four levels of sensitive area with $\sigma_b = 9.0$ microns, determined by analysis for four estimators and also the Cramér–Rao lower bound. Sensitive area given as percent of pixel area. The baseline value for the experimental configuration is 65%.

Method	Sensitive area (percent)			
	0%	25%	65%	100%
Binary centroid	260.0	254.5	246.0	245.0
Grayscale centroid	230.5	225.5	219.0	219.0
Ellipsoidal contour	23.0	23.5	24.0	24.0
Butterworth tepuy	23.0	23.0	22.0	21.0
Cramér–Rao lower bound	15.0	17.0	17.5	18.0

Study 8.7 shows the effects of pixel sensitive area and pixel geometry on landmark location uncertainty. Figure 8.10 shows that when the smoothing radius (blur) is approximately the dimension of a pixel, then the pixel sensitive area has a limited impact on location uncertainty.

Figure 8.11, in contrast, shows that when the smoothing radius is small relative to a pixel, then sensitive area has a significant impact on the model-based estimators, particularly the ellipsoidal contour algorithm. Over the range investigated, the pixel aspect ratio had minimal effect.

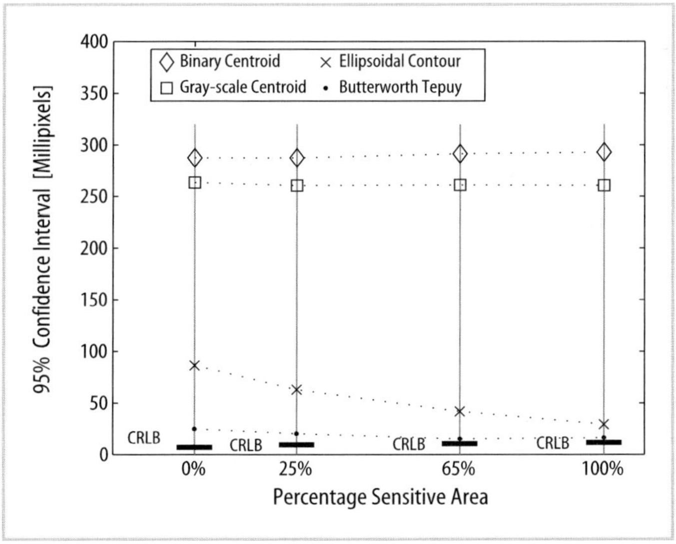

Figure 8.11. Landmark location uncertainty performance map: effects of changing the pixel sensitive area, $\sigma_b = 2.0$ microns

Table 8.9. Radius of 95% circular confidence interval for four levels of sensitive area with $\sigma_b = 2.0$ microns, determined by analysis for four estimators and also the Cramér–Rao lower bound. Sensitive area given as percent of pixel area. The baseline value for the experimental configuration is 65%.

Method	Sensitive area (percent)			
	0%	25%	65%	100%
Binary centroid	287.0	287.5	291.5	293.0
Grayscale centroid	263.5	260.5	261.0	260.5
Ellipsoidal contour	86.5	63.0	42.0	29.5
Butterworth tepuy	25.0	20.5	15.5	16.5
Cramér–Rao lower bound	7.0	9.5	10.5	11.5

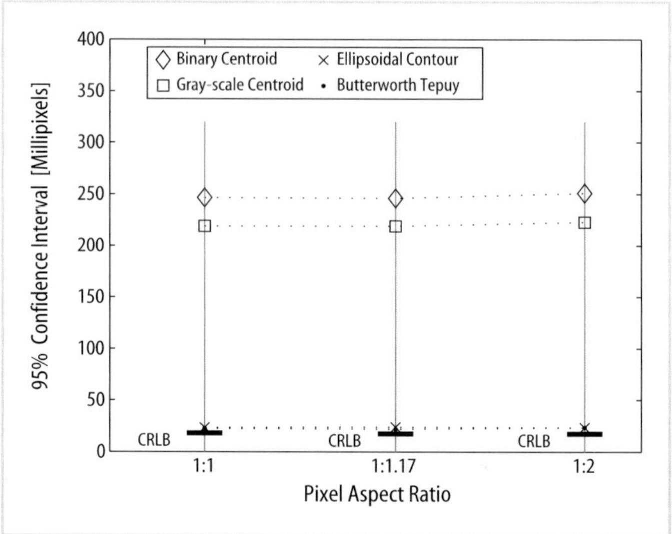

Figure 8.12. Landmark location uncertainty performance map: effects of different pixel aspect ratios

Table 8.10. Radius of 95% circular confidence interval for three values of the pixel aspect ratio, determined by analysis for four estimators and the Cramér–Rao lower bound. The pixel aspect ratio is the ratio of the smallest to the largest pixel dimensions. The baseline value for the experimental configuration is 1 : 1.17.

Method	Pixel aspect ratio		
	1:1	1:1.7	1:2
Binary centroid	246.5	246.0	251.0
Grayscale centroid	219.0	219.0	223.0
Ellipsoidal contour	23.5	24.0	23.5
Butterworth tepuy	22.5	22.0	23.5
Cramér–Rao lower bound	13.0	17.5	17.5

8.8 Study of the Effects of Nonuniform Illumination on Landmark Location Uncertainty

Figure 8.13 presents the results for the impact of nonuniform illumination on the 35-mm experimental configuration. In this case, a luminosity gradient was superimposed onto the simulated data. The data for Figure 8.13 are presented in Table 8.11.

In this study, the luminosity gradient is expressed as percent variation across the diameter of the landmark. The baseline illuminance is 30% of full scale for the imager, or 77 ADC counts of the eight-bit quantizer. The variations correspond to differences of zero, three, and six ADC counts of illuminance across the image of the landmark. Study 8.8 shows that the principal impact of nonuniform illumination is to introduce bias into the model-based estimators, which do not model variation in illumination. The data show that the centroid estimators are robust to nonuniform

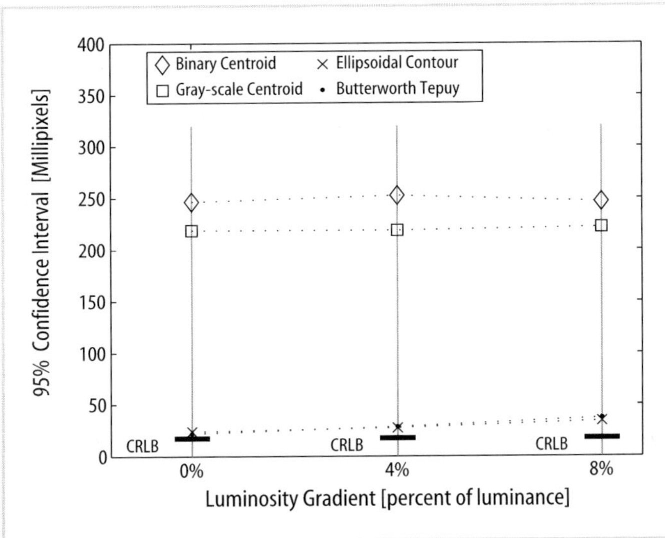

Figure 8.13. Landmark location uncertainty performance map: effects of nonuniform illumination

Table 8.11. Radius of 95% circular confidence interval for three values of nonuniform illumination, determined by analysis for four estimators and the Cramér–Rao lower bound. Nonuniform illumination is expressed in percent variation of luminance across the landmark. The baseline value for the experimental configuration is 0 nonuniform illumination.

Method	Nonuniform illumination		
	0.0	4.0%	8.0%
Binary centroid	246.5	252.5	246.5
Grayscale centroid	219.0	219.0	222.0
Ellipsoidal contour	24.0	27.5	34.0
Butterworth tepuy	22.0	28.5	37.0
Cramér–Rao lower bound	17.5	17.5	17.5

illumination. The Cramér–Rao lower bound depends on the sensitivity of the data to the displacement of the landmark, and is only affected by bias to the extent that the signal grows strong as the illumination is increased toward 100% of full scale. In spite of the marked difference between the models underlying the ellipsoidal contour and Butterworth tepuy estimators, they are very similarly impacted by nonuniform illumination. This study indicates that the performance of the model-based estimators can be improved by incorporating a representation of nonuniform illumination, for example as a two-element gradient vector.

8.9 Study of the Effects of Amplitude Quantization

Figure 8.14 presents the results for the impact of the amplitude quantizer resolution on the 35-mm experimental configuration. Quantizer resolution levels of

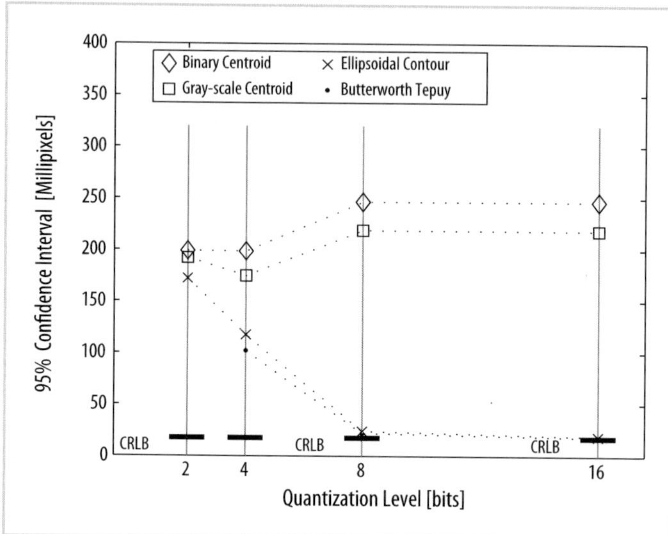

Figure 8.14. Landmark location uncertainty performance map: effects of amplitude quantization

Table 8.12. Radius of 95% circular confidence interval for four levels of amplitude quantization, determined by analysis for four estimators and also the Cramér–Rao lower bound. Quantization is expressed in ADC bits. The baseline value for the experimental configuration is $q = 8$ bits.

Method	Amplitude quantization			
	2	4	8	16
Binary centroid	199.0	198.5	246.5	246.5
Grayscale centroid	192.0	174.5	219.0	218.5
Ellipsoidal contour	172.0	117.5	24.0	19.0
Butterworth tepuy	–	102.0	22.0	18.0
Cramér–Rao lower bound	17.5	17.5	17.5	17.5

$q = \{2, 4, 8, 16\}$ bits are considered. A quantizer resolution of one bit is not considered because the grayscale centroid estimator is equivalent to the binary centroid estimator and the model-based estimators are ineffective.

As seen in Figure 2.14, locales grow large when the quantization resolution is coarse. This is why all estimators approach the same uncertainty for $q = 2$. The Cramér–Rao bound is calculated for analog intensity data, which do not include amplitude quantization, and so the CRLB uncertainty is the same in each of the four cases shown.

For the model-based estimators, the uncertainty drops dramatically with an increase from four to eight bits of signal. Since the estimators with eight-bit quantization approach the Cramér–Rao lower bound, only a slight improvement in model-based estimator performance is achieved by increasing the quantizer resolution to 16 bits. Recall that the baseline case has an imager noise of 1/256 full scale, or one count of eight-bit quantization. A lower noise level is expected to lead to an improvement.

At low resolution, most of the uncertainty is due to the locale phenomenon. For the baseline configuration considered in this study, the locale phenomenon is less important at a quantization level of eight bits, and the imager noise becomes the main uncertainty factor.

9 Conclusions

The accuracy of image-based measurement depends on the accuracy with which the locations of landmarks are determined. This work has advanced the state-of-the-art in precision landmark location in the following ways:

- The confidence interval has been introduced as a measure of landmark location uncertainty because it is able to accommodate the nonstationary statistics of the problem
- An extended model of the physics of image formation has been developed, which captures aspects of the physical process that contribute to the determination of landmark location uncertainty
- The Cramér–Rao lower bound for landmark location uncertainty has been determined, establishing a minimum level of uncertainty for any unbiased estimator, independent of the estimation algorithm used
- A set of tools for analyzing the uncertainties of practical landmark and imaging configurations and location estimation algorithms have been created
- Two new, model-based landmark location estimators have been demonstrated, which perform near the Cramér–Rao theoretical limit
- These tools have been validated by developing and performing an experiment that can directly and accurately measure the error in landmark location estimates.

There are many ways to process an image and estimate the location of a landmark. The Cramér–Rao lower bound has been applied to establish the minimum uncertainty that any estimator can achieve, given the data and noise present in an image. To carry out the CRLB calculation, it has been necessary to augment the model of image formation with information about the sensitive area of each pixel and a model that smoothes as an image is formed. When locations are estimated with uncertainties of a small fraction of a pixel, these characteristics play a significant role. Additionally, smoothing eliminates discontinuities in the illuminance of the imager, rendering the analysis mathematically tractable.

The EyeBox tool kit developed for this study provides a set of tools for analyzing specific landmark, imaging, and estimator configurations. The tool kit has functions for computing the CRLB or analyzing the performances of practical landmark location estimators. One challenge when studying precision landmark location, which is reflected in the EyeBox tool kit, is that precision depends on many details of configuration. At the limit, where uncertainty is measured in tens of millipixels, the location uncertainty may change by a factor of two when a lens aperture is changed from $\#f/3$ to $\#f/6$. The strength of the tool kit is that it can analyze a specific imager/landmark configuration, while the weakness is that it must analyze a specific configuration. Nonetheless, by studying the performance of specific configurations, systems can be engineered to achieve their maximum performance.

Inspired by the accuracy of the model-based Nalwa–Binford algorithm for locating edges, two model-based landmark location estimation algorithms have been developed. Somewhat to our surprise, these estimators give performances approaching the Cramér–Rao lower bound. The ellipsoidal contour algorithm, in particular, only requires that edge pixels of the landmark image are processed, and has proven to be a precise and computationally efficient tool.

To validate the EyeBox tool, as well as the Cramér–Rao lower bound, an experiment has been performed to directly measure landmark location precision. It is straightforward to estimate the location of the image of a landmark. It is more challenging to measure the estimator error, since the latter requires knowledge of where the landmark truly lies. To meet this challenge, a high-precision artwork and an algorithm have been designed that are sufficiently accurate to be used as a reference. The tool, called the nano-triangles estimator, averages over hundreds of pixels, imaging artwork that is laid out in order to cancel first-order error sources and to provide a consistency check on its own performance. The nano-triangles experiment confirmed the analytic predictions, although at a somewhat higher level of uncertainty, indicating that there is an as-yet unmodeled error source in the landmark location process.

The results of this work answer several questions posed by investigators during 30 years of research into landmark location in digital images. As the use of images for measurement grows, the methods and results presented here will provide a foundation for the study of diverse imaging, landmark, and estimator configurations, new and more complete models of the image formation process, and the engineering of systems to achieve ultimate performance.

Appendix A
List of Symbols

Symbol	Quantity
$*$	2-D convolution operator
c_p	adjusted focal length
$E({}^ix, {}^iy)$	illuminance function; parameters in image coordinates
$f({}^Lx, {}^Ly)$	landmark luminosity function; parameters in landmark coordinates
$G_f\left({}^c\mathcal{P}_{\underset{L}{o}}\right)$	landmark description transformation
$H({}^ix, {}^iy)$	optical flux function; parameters in image coordinates
$I(i,j)$	digital image function; parameters in pixel coordinates
J	Fisher information matrix
J_1	Bessel function of the first kind and order 1
$J(i,j)$	discrete analog intensity function; parameters in pixel coordinates
K_o	radial lens distortion coefficient: linear term
K_1	radial lens distortion coefficient: cubic term
k	propagation number
k_x	density of pixels in the imager: x direction (in pixels/mm)
k_y	density of pixels in the imager: y direction (in pixels/mm)
P_1	decentering lens distortion coefficient: x direction
P_2	decentering lens distortion coefficient: y direction
${}^wP$	point in world coordinates $\in \Re^3$
${}^cP$	point in camera coordinates $\in \Re^3$
${}^LP$	point in landmark coordinates $\in \Re^2$
${}^iP$	point in image coordinates $\in \Re^2$
${}^pP$	point in pixel coordinates $\in \Re^2$
Q_i	amplitude quantization transformation
Q_s	spatial sampling transformation
r_a	Airy disk radius
$T(I)$	estimator of the center of the landmark
Φ	geometric optics transfer function
Ξ	diffraction transfer function
λ	wavelength
σ_b	standard deviation of the Gaussian smoothing function
ω	pitch
κ	roll
φ	yaw

Appendix B
Glossary

Aberration: The failure of an optical lens to produce an exact point-to-point correspondence between the object and its resulting image.

ADC: Analog-to-digital converter, used to convert the charge collected in a pixel of an imager to a digital value.

Adjusted focal length: Distance along the principal ray from imager to the principal point of the lens, also called the principal distance, c_p.

Airy pattern: Diffraction pattern caused by a circular aperture.

Bundle adjustment: an estimation process used in photogrammetry to simultaneously estimate the locations of target points (landmarks) and the pose of one or several camera stations.

Centroid algorithm: A landmark location estimator that operates forming the mathematical centroid of the image region identified as the landmark (see Equation (2.22)).

CRLB: Cramér–Rao lower bound; the lower bound on uncertainty of an unbiased estimate based on the sensitivity of the data to changes in the estimated quantity, and not the noise present in the data.

Cubature formula: Formula used for the numerical computation of a multiple integral. The 2-D Simpson's rule is an example of a cubature formula.

DOF: Degrees of freedom, for example an unconstrained body has six degrees of motion freedom, or 6 DOF.

Diffraction: Dispersion of light that occurs when light waves pass through a barrier, such as the aperture of a camera. Diffraction has the effect of applying a spatial low-pass filter to the optical signal.

Digital image function: Function describing the discrete analog intensity function after amplitude quantization.

Discrete analog intensity function: Function describing the illuminance function after spatial quantization.

Fiducial marks: reference points in a scene that can be used to locate objects or calibrate measurements.

Fill factor: The percentage of photosensitive area within a pixel boundary.

Focal point: A point from which a portion of spherical waves converges or diverges.

Gray level: A quantized measurement of the discrete analog intensity function, typically in the range between pure white and black.

Grayscale: Variations in values from white through shades of gray to black in a digitized image, where black is assigned a value of zero and white a value of one.

Grayscale image: An image consisting of an array of pixels whose intensities can take more than two values.

Illuminance: The intensity of the light incident on a surface (in lux). Digital imager sensitivity is often specified in counts/lux-second. (See also *luminance*).

Illuminance function: function representing the distribution of optical flux on the imager, including consideration of diffraction.

Image capture: The process of acquiring an image of a scene, from sensor irradiation to acquisition of a digital image.

Image distortion: A situation in which the image is not exactly true to scale to the object scale.

Image formation: Generation of an image of an object or scene on the imaging sensor. It includes effects from the optics, illumination and the imager itself.

Image plane: The plane surface of the imaging sensor perpendicular to the principal ray.

Image processing: Digital manipulation of an image to aid feature visibility, make measurements or to alter image contents.

Landmark description transformation: Transformation that creates the landmark luminosity function from the pose of its center.

Landmark luminosity function: Function that describes the landmark geometry on the landmark plane.

Locale: Equivalence class of the landmark location.

Luminance: The measure of intensity of light radiated from a surface in a particular direction, [candela/meter^2]. (See also *illuminance*).

Machine vision: The use of devices for optical noncontact sensing to automatically receive and interpret an image of a real scene, in order to obtain information and/or control machines or processes.

MTF: Modulation transfer function, a description of the ability of an optical system to transfer contrast at a given spatial frequency (*e.g.*, cycles per millimeter on a target at a standard distance). Lens systems generally form low-pass filters, and so higher spatial frequencies are attenuated.

Optical flux function: Function representing the spatial distribution of the landmark's luminosity after transformation via geometrical optics, but without considering diffraction smoothing. Examples that contrast the optical flux function and illuminance function can be seen in Figures 2.1 and 2.3.

Photogrammetry: Science and art of taking measurements from images.

Pixel: An area element of an image. In a monochrome digital image, each pixel corresponds to a number that is the intensity (illuminance) on the pixel's sensitive area.

Principal distance: See *adjusted focal length.*

Principal point, image: Point where the principal ray intersects the image plane, the origin of image coordinates.

Principal point, lens: Origin of the camera coordinate frame and location of the pinhole in a pinhole camera model for geometric optics.

Principal ray: Optical axis of the camera lens, which passes through the principal points of the lens and imager.

Raster scan: A scanning pattern, generally performed from left to right while progressing from top to bottom of the imaging sensor.

Scene: A portion of space imaged by a vision system.

Sensitive area: The region of a pixel that is sensitive to incident light.

Appendix C
Error Estimate of the Noncollocated 2-D Simpson Rule

The 2-D numerical integration developed in Chapter 5 represents an approximation to the real value of the integral. An estimate of the error caused by this approximation is required in order to ensure a grid design that will produce sufficiently small inaccuracies during the analysis of landmark location experiments. Several results are needed to find an expression for the approximation error $\varepsilon(f)$. A roadmap of the theorems required to formulate a bound on $\varepsilon(f)$ is shown in Figure C.1.

Theorem 1.
The integration formula in Equation (5.23) *is exact* ($\varepsilon(f) = 0$) *if* $f(x, y)$ *is given by any second-order polynomial in* $\Re^2$.

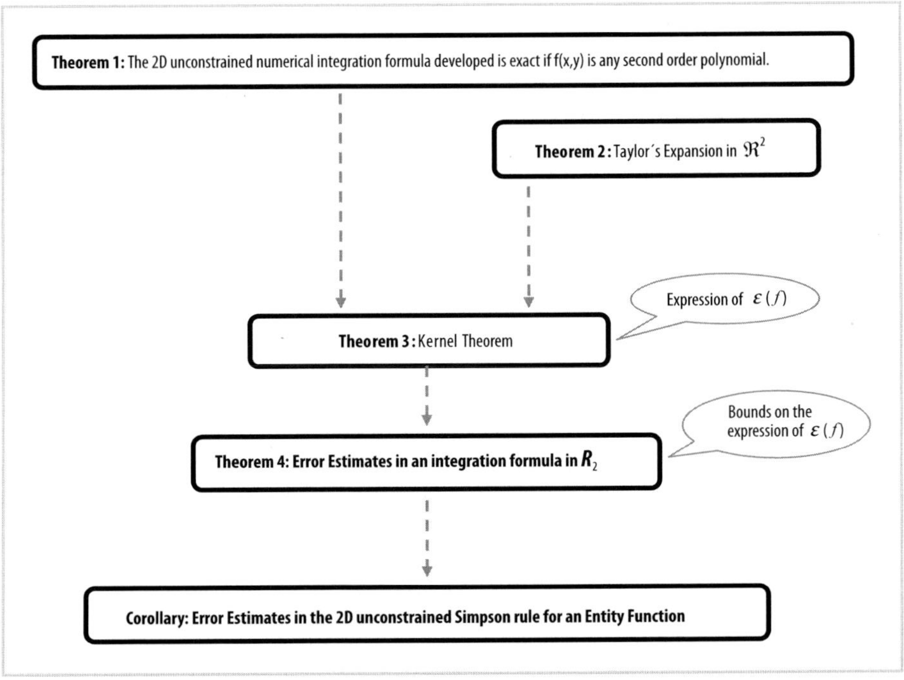

Figure C.1. Roadmap of theorems needed to formulate the 2-D noncollocated numerical integration error bound

Proof. Given the second-order polynomial $f(x,y) = Ax^2 + Bx + Cxy + Dy^2 + Ey + F$, the analytical integration formula in the interval $\left[x_o + \delta_x^L, x_{n_x} - \delta_x^U\right] \times \left[y_o + \delta_y^L, y_{n_y} - \delta_y^U\right]$ is

$$\int_{x_o+\delta_x^L}^{x_{n_x}-\delta_x^U} \int_{y_o+\delta_y^L}^{y_{n_y}-\delta_y^U} f(x,y)\,dy\,dx$$

$$= \int_{x_o+\delta_x^L}^{x_{n_x}-\delta_x^U} \left[Ax^2y + Bxy + Cx\frac{y^2}{2} + D\frac{y^3}{3} + E\frac{y^2}{2} + Fy\right]_{y_o+\delta_y^L}^{y_{n_y}-\delta_y^U} dx$$

$$\begin{aligned}
&= \frac{A}{3}\left(\left(y_{n_y} - \delta_y^U\right) - \left(y_o + \delta_y^L\right)\right)\left(\left(x_{n_x} - \delta_x^U\right)^3 - \left(x_o + \delta_x^L\right)^3\right)\\
&+ \frac{B}{2}\left(\left(y_{n_y} - \delta_y^U\right) - \left(y_o + \delta_y^L\right)\right)\left(\left(x_{n_x} - \delta_x^U\right)^2 - \left(x_o + \delta_x^L\right)^2\right)\\
&+ \frac{C}{4}\left(\left(y_{n_y} - \delta_y^U\right)^2 - \left(y_o + \delta_y^L\right)^2\right)\left(\left(x_{n_x} - \delta_x^U\right)^2 - \left(x_o + \delta_x^L\right)^2\right)\\
&+ \frac{D}{3}\left(\left(y_{n_y} - \delta_y^U\right)^3 - \left(y_o + \delta_y^L\right)^3\right)\left(\left(x_{n_x} - \delta_x^U\right) - \left(x_o + \delta_x^L\right)\right)\\
&+ \frac{E}{2}\left(\left(y_{n_y} - \delta_y^U\right)^2 - \left(y_o + \delta_y^L\right)^2\right)\left(\left(x_{n_x} - \delta_x^U\right) - \left(x_o + \delta_x^L\right)\right)\\
&+ F\left(\left(y_{n_y} - \delta_y^U\right) - \left(y_o + \delta_y^L\right)\right)\left(\left(x_{n_x} - \delta_x^U\right) - \left(x_o + \delta_x^L\right)\right) . \qquad \text{(C.1)}
\end{aligned}$$

We will prove that the integration formula in Equation (5.23) is exact for the second-order polynomial $f(x,y)$ for the case of $n_x = n_y = 2$. The proof expands to any grid of $[n_x + 1] \times \left[n_y + 1\right]$ knots (according to Equations (5.18) and (5.21), n_x and n_y are even numbers), since any square integration interval can be decomposed into a tessellation of 3×3 knots.

Equation (5.23) can be written as:

$$\int_{y_o+\delta_y^L}^{y_f-\delta_y^U} \int_{x_o+\delta_x^L}^{x_f-\delta_x^U} f(x,y)\,dx\,dy =$$

$$\frac{h_x h_y}{9}\sum_{i=0}^{2}\sum_{j=0}^{2}\left(S_{i,j} - R_{i,j}(h_x, h_y, \delta_x^L, \delta_x^H, \delta_y^L, \delta_y^H)\right) f(x_i, y_j) + \varepsilon(f) , \qquad \text{(C.2)}$$

where

$$S = \begin{bmatrix} 1 & 4 & 1 \\ 4 & 16 & 4 \\ 1 & 4 & 1 \end{bmatrix}$$

$$R = \begin{bmatrix} V_1^{xL} + V_3^{xU} + V_1^{yL} + V_3^{yU} & V_2^{xL} + V_2^{xU} + 4V_1^{yL} + 4V_3^{yU} & V_3^{xL} + V_1^{xU} + V_1^{yL} + V_3^{yU} \\ 4V_1^{xL} + 4V_3^{xU} + V_2^{yL} + V_2^{yU} & 4V_2^{xL} + 4V_2^{xU} + 4V_2^{yL} + 4V_2^{yU} & 4V_3^{xL} + 4V_1^{xU} + V_2^{yL} + V_2^{yU} \\ V_1^{xL} + V_3^{xU} + V_3^{yL} + V_1^{yU} & V_2^{xL} + V_2^{xU} + 4V_3^{yL} + 4V_1^{yU} & V_3^{xL} + V_1^{xU} + V_3^{yL} + V_1^{yU} \end{bmatrix}$$
$$- \begin{bmatrix} V_1^{xL}V_1^{yL} + V_3^{xU}V_1^{yL} + V_1^{xL}V_3^{yU} + V_3^{xU}V_3^{yU} & V_2^{xL}V_1^{yL} + V_2^{xU}V_1^{yL} + V_2^{xL}V_3^{yU} + V_2^{xU}V_3^{yU} & V_3^{xL}V_1^{yL} + V_1^{xU}V_1^{yL} + V_3^{xL}V_3^{yU} + V_1^{xU}V_3^{yU} \\ V_1^{xL}V_2^{yL} + V_3^{xU}V_2^{yL} + V_1^{xL}V_2^{yU} + V_3^{xU}V_2^{yU} & V_2^{xL}V_2^{yL} + V_2^{xU}V_2^{yL} + V_2^{xL}V_2^{yU} + V_2^{xU}V_2^{yU} & V_3^{xL}V_2^{yL} + V_1^{xU}V_2^{yL} + V_3^{xL}V_2^{yU} + V_1^{xU}V_2^{yU} \\ V_1^{xL}V_3^{yL} + V_3^{xU}V_3^{yL} + V_1^{xL}V_1^{yU} + V_3^{xU}V_1^{yU} & V_2^{xL}V_3^{yL} + V_2^{xU}V_3^{yL} + V_2^{xL}V_1^{yU} + V_2^{xU}V_1^{yU} & V_3^{xL}V_3^{yL} + V_1^{xU}V_3^{yL} + V_3^{xL}V_1^{yU} + V_1^{xU}V_1^{yU} \end{bmatrix},$$

and

$$
\begin{aligned}
f(x_i, y_j) &= Ax_i^2 + Bx_i + Cx_iy_j + Dy_j^2 + Ey_j + F \\
&= A\left(\frac{(2-i)x_o + ix_2}{2}\right)^2 + B\left(\frac{(2-i)x_o + ix_2}{2}\right) \\
&\quad + C\left(\frac{(2-i)x_o + ix_2}{2}\right)\left(\frac{(2-j)y_o + jy_2}{2}\right) \\
&\quad + D\left(\frac{(2-j)y_o + jy_2}{2}\right)^2 + E\left(\frac{(2-j)y_o + jy_2}{2}\right) + F \qquad i,j \in \{0,1,2\} \ .
\end{aligned}
\tag{C.3}
$$

After algebraic manipulation we obtain:

$$
\begin{aligned}
&\int_{y_o+\delta_y^L}^{y_f-\delta_y^U} \int_{x_o+\delta_x^L}^{x_f-\delta_x^U} f(x,y)\,dx\,dy = \\
&\frac{A}{3}\left(x_o + \delta_x^L - x_2 + \delta_x^U\right)\begin{pmatrix} x_o^2 + 2\delta_x^L x_o + \left(\delta_x^L\right)^2 + x_o x_2 - \delta_x^U x_o \\ +\delta_x^L x_2 - \delta_x^L\delta_x^U + x_2^2 - 2\delta_x^U x_2 + \left(\delta_x^U\right)^2 \end{pmatrix}\left(y_o + \delta_y^L - y_2 + \delta_y^U\right) \\
&+ \frac{B}{2}\left(x_o + \delta_x^L - x_2 + \delta_x^U\right)\left(x_o + \delta_x^L + x_2 - \delta_x^U\right)\left(y_o + \delta_y^L - y_2 + \delta_y^U\right) \\
&+ \frac{C}{4}\left(x_o + \delta_x^L - x_2 + \delta_x^U\right)\left(x_o + \delta_x^L + x_2 - \delta_x^U\right)\left(y_o + \delta_y^L - y_2 + \delta_y^U\right)\left(y_o + \delta_y^L + y_2 - \delta_y^U\right) \\
&+ \frac{D}{3}\left(y_o + \delta_y^L - y_2 + \delta_y^U\right)\begin{pmatrix} y_o^2 + 2\delta_y^L y_o + \left(\delta_y^L\right)^2 + y_o y_2 - \delta_y^U y_o \\ +\delta_y^L y_2 - \delta_y^L\delta_y^U + y_2^2 - 2\delta_y^U y_2 + \left(\delta_y^U\right)^2 \end{pmatrix}\left(x_o + \delta_x^L - x_2 + \delta_x^U\right) \\
&+ \frac{E}{2}\left(y_o + \delta_y^L - y_2 + \delta_y^U\right)\left(y_o + \delta_y^L + y_2 - \delta_y^U\right)\left(x_o + \delta_x^L - x_2 + \delta_x^U\right) \\
&+ F\left(x_o + \delta_x^L - x_2 + \delta_x^U\right)\left(y_o + \delta_y^L - y_2 + \delta_y^U\right) \\
&+ \varepsilon(f) \\
&\Rightarrow \int_{y_o+\delta_y^L}^{y_f-\delta_y^U} \int_{x_o+\delta_x^L}^{x_f-\delta_x^U} f(x,y)\,dx\,dy = \\
&\frac{A}{3}\left(\left(y_2 - \delta_y^U\right) - \left(y_o + \delta_y^L\right)\right)\left(\left(x_2 - \delta_x^U\right)^3 - \left(x_o + \delta_x^L\right)^3\right) \\
&+ \frac{B}{2}\left(\left(y_2 - \delta_y^U\right) - \left(y_o + \delta_y^L\right)\right)\left(\left(x_2 - \delta_x^U\right)^2 - \left(x_o + \delta_x^L\right)^2\right) \\
&+ \frac{C}{4}\left(\left(y_2 - \delta_y^U\right)^2 - \left(y_o + \delta_y^L\right)^2\right)\left(\left(x_2 - \delta_x^U\right)^2 - \left(x_o + \delta_x^L\right)^2\right) \\
&+ \frac{D}{3}\left(\left(y_2 - \delta_y^U\right)^3 - \left(y_o + \delta_y^L\right)^3\right)\left(\left(x_2 - \delta_x^U\right) - \left(x_o + \delta_x^L\right)\right) \\
&+ \frac{E}{2}\left(\left(y_2 - \delta_y^U\right)^2 - \left(y_o + \delta_y^L\right)^2\right)\left(\left(x_2 - \delta_x^U\right) - \left(x_o + \delta_x^L\right)\right) \\
&+ F\left(\left(y_2 - \delta_y^U\right) - \left(y_o + \delta_y^L\right)\right)\left(\left(x_2 - \delta_x^U\right) - \left(x_o + \delta_x^L\right)\right) \\
&+ \varepsilon(f) \ .
\end{aligned}
\tag{C.4}
$$

Since Equation (C.1) is equal to Equation (C.4) for $n_x = n_y = 2$, this implies that $\varepsilon(f) = 0$.

Theorem 1 is proven.

As for the 1-D case, Theorem 1 cannot be generalized for third-order polynomials in $\Re^2$ for the case where the integration interval is not coincidental with the integration grid.

Definition 1. *Let $R_2 \in \Re^2$ be a rectangle $a \le x \le b$, $c \le y \le d$.*

Definition 2. *The derivatives of $f(x, y)$ will be written as follows:*

$$f^{(i,j)}(x,y) \equiv \frac{\partial^{i+j}}{\partial x^i \partial y^j} f(x,y) \qquad i \ge 0,\ j \ge 0$$
$$f^{(0,0)}(x,y) \equiv f(x,y)$$
$$f^{(i,j)}(a,c) \equiv f^{(i,j)}(x,y)\big|_{(x,y)=(a,c)}$$

Definition 3. *The selector function $\psi(\alpha, \beta, \gamma)$ is defined as:*

$$\psi(\alpha,\beta,\gamma) = \begin{cases} 1 & \alpha \le \beta < \gamma \\ -1 & \gamma \le \beta < \alpha \\ 0 & otherwise \end{cases} \tag{C.5}$$

Theorem 2 (Taylor's Series Expansion in $\Re^2$ (from Stroud 1971)). *If $f(x, y)$ has the property that all the derivatives*

$$f^{(i,j)}(x,y) \quad 0 \le i \le p \quad 0 \le j \le q \quad p \ge 1 \quad q \ge 1$$

are defined and continuous in R_2, then $f(x, y)$ has the expansion

$$\begin{aligned} f(x,y) = {} & \sum_{i=0}^{p-1}\sum_{j=0}^{q-1} \frac{(x-a_o)^i}{i!}\frac{(y-c_o)^j}{j!} f^{(i,j)}(a_o,c_o) \\ & + \sum_{j=0}^{q-1} \frac{(y-c_o)^j}{j!} \int_{a_o}^{x} \frac{(x-u)^{p-1}}{(p-1)!} f^{(p,j)}(u,c_o)\,du \\ & + \sum_{i=0}^{p-1} \frac{(x-a_o)^i}{i!} \int_{c_o}^{y} \frac{(x-v)^{q-1}}{(q-1)!} f^{(i,q)}(a_o,v)\,dv \\ & + \int_{a_o}^{x}\int_{c_o}^{y} \frac{(x-u)^{p-1}}{(p-1)!}\frac{(y-v)^{q-1}}{(q-1)!} f^{(p,q)}(u,v)\,du\,dv\,, \end{aligned} \tag{C.6}$$

where (a_o, c_o) is any point of R_2.

The proof of Theorem 2 is given in Chapter 5 of Stroud (1971).

Definition 4. $B_{p,q}$ *is the set (space) of all functions* $f(x,y)$ *with the following properties:*

i. The derivative $f^{(p,q)}(x,y)$ *is Riemann-integrable on* R_2
ii. The derivatives $f^{(p,j)}(x,c_o), j<q$, *are Riemann-integrable on* $a \le x \le b$.
iii. The derivatives $f^{(i,q)}(a_o,y), i<p$, *are Riemann-integrable on* $c \le y \le d$.
iv. The derivatives $f^{(i,j)}(a_o,c_o), i<p, j<q$ *exist*
v. Taylor's expansion (Equation (C.6)) holds for $f(x,y)$ *for all* (x,y) *in* R_2

Theorem 3 (Kernel Theorem (from Stroud 1971)).
Given an integration formula of the form

$$\int\limits_{y_o+\delta_y^L}^{y_f-\delta_y^U} \int\limits_{x_o+\delta_x^L}^{x_f-\delta_x^U} f(x,y)\,dx\,dy = \sum_k A_k f(x_k,y_k) + \varepsilon(f)\,, \tag{C.7}$$

which is exact, in other words $\varepsilon(f)=0$ *when* $f(x,y)$ *is a polynomial of degree* $<p$ *in x and degree* $<q$ *in y, then there exist kernel functions:*

$$\begin{array}{lll} K_{p,q}(x,y,a_o,c_o) & (x,y)\in R_2 & (a_o,c_o)\in R_2 \\ K_{p,j}(x,a_o,c_o) & x\in[a,b] & j<q \\ K_{i,q}(y,a_o,c_o) & y\in[c,d] & i<p\,, \end{array} \tag{C.8}$$

so that

$$\begin{aligned} \varepsilon(f) = {} & \sum_{j=0}^{q-1} \int_a^b K_{p,j}(u,a_o,c_o) f^{(p,j)}(u,c_o)\,du \\ & + \sum_{i=0}^{p-1} \int_a^b K_{i,q}(v,a_o,c_o) f^{(i,q)}(a_o,v)\,dv \\ & + \int_a^b \int_c^d K_{p,q}(u,v,a_o,c_o) f^{(p,q)}(u,v)\,du\,dv \end{aligned} \tag{C.9}$$

whenever $f(x,y)\in B_{p,q}(a_o,c_o)$. *Furthermore, the kernel functions in Equation* (C.8) *can be defined by*

$$K_{p,j}(u,a_o,c_o) = \varepsilon\left[\frac{(x-u)^{p-1}}{(p-1)!}\psi(a_o,u,x)\frac{(y-c_o)^j}{j!}\right], \qquad u\in[a,b] \quad 0\le j<p \tag{C.10}$$

$$K_{i,q}(v,a_o,c_o) = \varepsilon\left[\frac{(x-a_o)^i}{i!}\frac{(y-v)^{q-1}}{(q-1)!}\psi(c_o,v,y)\right], \qquad v\in[c,d] \quad 0\le i<q \tag{C.11}$$

$$K_{p,q}(u,v,a_o,c_o) = \varepsilon\left[\frac{(x-u)^{p-1}}{(p-1)!}\psi(a_o,u,x)\frac{(y-v)^{q-1}}{(q-1)!}\psi(c_o,v,y)\right] (u,v)\in R_2\,. \tag{C.12}$$

Proof.[1] The 2-D Taylor series expansion shown in Equation (C.6) has the form:

$$f(x,y) = U_{p-1,q-1}(x,y) + W_{p,q}(x,y) \,, \tag{C.13}$$

where $U_{p-1,q-1}(x,y)$ is a polynomial of degree $< p$ in x and degree $< q$ in y, and $W_{p,q}(x,y)$ represents the rest of the elements of the expansion; that is:

$$U_{p-1,q-1}(x,y) = \sum_{i=0}^{p-1}\sum_{j=0}^{q-1} \frac{(x-a_o)^i}{i!}\frac{(y-c_o)^j}{j!} f^{(i,j)}(a_o,c_o) \tag{C.14}$$

$$\begin{aligned} W_{p,q}(x,y) &= \sum_{j=0}^{q-1} \frac{(y-c_o)^j}{j!} \int_{a_o}^{x} \frac{(x-u)^{p-1}}{(p-1)!} f^{(p,j)}(u,c_o)\,du \\ &+ \sum_{i=0}^{p-1} \frac{(x-a_o)^i}{i!} \int_{c_o}^{y} \frac{(x-v)^{q-1}}{(q-1)!} f^{(i,q)}(a_o,v)\,dv \\ &+ \int_{a_o}^{x}\int_{c_o}^{y} \frac{(x-u)^{p-1}}{(p-1)!}\frac{(y-v)^{q-1}}{(q-1)!} f^{(p,q)}(u,v)\,du\,dv \,. \end{aligned} \tag{C.15}$$

From Equations (C.7) and (C.13), we can write:

$$\begin{aligned} &\int_{y_o+\delta_y^L}^{y_f-\delta_y^U}\int_{x_o+\delta_x^L}^{x_f-\delta_x^U} f(x,y)\,dx\,dy \\ &= \int_{y_o+\delta_y^L}^{y_f-\delta_y^U}\int_{x_o+\delta_x^L}^{x_f-\delta_x^U} U_{p-1,q-1}(x,y)\,dx\,dy + \int_{y_o+\delta_y^L}^{y_f-\delta_y^U}\int_{x_o+\delta_x^L}^{x_f-\delta_x^U} W_{p,q}(x,y)\,dx\,dy \\ &= \sum_k A_k U_{p-1,q-1}(x_k,y_k) + \sum_k A_k W_{p,q}(x_k,y_k) + \varepsilon(f) \,, \end{aligned}$$

but since $U_{p-1.q-1}(x,y)$ is exact for degree $< p$ in x and degree $< q$ in y, this implies that:

$$\int_{y_o+\delta_y^L}^{y_f-\delta_y^U}\int_{x_o+\delta_x^L}^{x_f-\delta_x^U} W_{p,q}(x,y)\,dx\,dy = \sum_k A_k W_{p,q}(x_k,y_k) + \varepsilon(f)$$

or

$$\varepsilon(f) = \int_{y_o+\delta_y^L}^{y_f-\delta_y^U}\int_{x_o+\delta_x^L}^{x_f-\delta_x^U} W_{p,q}(x,y)\,dx\,dy - \sum_k A_k W_{p,q}(x_k,y_k) \,. \tag{C.16}$$

[1] This development is generalization of the proof presented in Stroud (1971).

Substituting the expression for $W_{p,q}(x,y)$ into Equation (C.16) and rearranging terms, we obtain:

$$
\begin{aligned}
\varepsilon(f) = & \int_{y_o+\delta_y^L}^{y_f-\delta_y^U} \int_{x_o+\delta_x^L}^{x_f-\delta_x^U} \left[\sum_{j=0}^{q-1} \frac{(y-c_o)^j}{j!} \int_{a_o}^{x} \frac{(x-u)^{p-1}}{(p-1)!} f^{(p,j)}(u,c_o)\,du \right] dx\,dy \\
& - \sum_k A_k \left[\sum_{j=0}^{q-1} \frac{(y_k-c_o)^j}{j!} \int_{a_o}^{x} \frac{(x-u)^{p-1}}{(p-1)!} f^{(p,j)}(u,c_o)\,du \right]_k \\
& + \int_{y_o+\delta_y^L}^{y_f-\delta_y^U} \int_{x_o+\delta_x^L}^{x_f-\delta_x^U} \left[\sum_{i=0}^{p-1} \frac{(x-a_o)^i}{i!} \int_{c_o}^{y} \frac{(x-v)^{q-1}}{(q-1)!} f^{(i,q)}(a_o,v)\,dv \right] dx\,dy \\
& - \sum_k A_k \left[\sum_{i=0}^{p-1} \frac{(x_k-a_o)^i}{i!} \int_{c_o}^{y} \frac{(y-v)^{q-1}}{(q-1)!} f^{(i,q)}(a_o,v)\,dv \right]_k \\
& + \int_{y_o+\delta_y^L}^{y_f-\delta_y^U} \int_{x_o+\delta_x^L}^{x_f-\delta_x^U} \left[\int_{a_o}^{x} \int_{c_o}^{y} \frac{(x-u)^{p-1}}{(p-1)!} \frac{(y-v)^{q-1}}{(q-1)!} f^{(p,q)}(u,v)\,du\,dv \right] dx\,dy \\
& - \sum_k A_k \left[\int_{a_o}^{x} \int_{c_o}^{y} \frac{(x-u)^{p-1}}{(p-1)!} \frac{(y-v)^{q-1}}{(q-1)!} f^{(p,q)}(u,v)\,du\,dv \right]_k .
\end{aligned}
$$

Bringing out the summation of the i and j terms, we get:

$$
\begin{aligned}
\varepsilon(f) = & \sum_{j=0}^{q-1} \int_{y_o+\delta_y^L}^{y_f-\delta_y^U} \int_{x_o+\delta_x^L}^{x_f-\delta_x^U} \left[\int_{a_o}^{x} \frac{(y-c_o)^j}{j!} \frac{(x-u)^{p-1}}{(p-1)!} f^{(p,j)}(u,c_o)\,du \right] dx\,dy \\
& - \sum_{j=0}^{q-1} \sum_k A_k \left[\int_{a_o}^{x} \frac{(y_k-c_o)^j}{j!} \frac{(x_k-u)^{p-1}}{(p-1)!} f^{(p,j)}(u,c_o)\,du \right]_k \\
& + \sum_{i=0}^{p-1} \int_{y_o+\delta_y^L}^{y_f-\delta_y^U} \int_{x_o+\delta_x^L}^{x_f-\delta_x^U} \left[\int_{c_o}^{y} \frac{(x-a_o)^i}{i!} \frac{(x-v)^{q-1}}{(q-1)!} f^{(i,q)}(a_o,v)\,dv \right] dx\,dy \\
& - \sum_{i=0}^{p-1} \sum_k A_k \left[\int_{c_o}^{y} \frac{(x_k-a_o)^i}{i!} \frac{(y_k-v)^{q-1}}{(q-1)!} f^{(i,q)}(a_o,v)\,dv \right]_k
\end{aligned}
$$

$$+\int_{y_o+\delta_y^L}^{y_f-\delta_y^U}\int_{x_o+\delta_x^L}^{x_f-\delta_x^U}\left[\int_{a_o}^{x}\int_{c_o}^{y}\frac{(x-u)^{p-1}}{(p-1)!}\frac{(y-v)^{q-1}}{(q-1)!}f^{(p,q)}(u,v)\,du\,dv\right]dx\,dy$$

$$-\sum_k A_k\left[\int_{a_o}^{x}\int_{c_o}^{y}\frac{(x_k-u)^{p-1}}{(p-1)!}\frac{(y_k-v)^{q-1}}{(q-1)!}f^{(p,q)}(u,v)\,du\,dv\right]_k. \quad \text{(C.17)}$$

Using selector function $\psi(\alpha,\beta,\gamma)$, Equation (C.17) can be written as:

$$\varepsilon(f)=\sum_{j=0}^{q-1}\int_{y_o+\delta_y^L}^{y_f-\delta_y^U}\int_{x_o+\delta_x^L}^{x_f-\delta_x^U}\left[\int_a^b\frac{(y-c_o)^j}{j!}\psi(a_o,u,x)\frac{(x-u)^{p-1}}{(p-1)!}f^{(p,j)}(u,c_o)\,du\right]dx\,dy$$

$$-\sum_{j=0}^{q-1}\sum_k A_k\left[\int_a^b\frac{(y_k-c_o)^j}{j!}\psi(a_o,u,x)\frac{(x_k-u)^{p-1}}{(p-1)!}f^{(p,j)}(u,c_o)\,du\right]_k$$

$$+\sum_{i=0}^{p-1}\int_{y_o+\delta_y^L}^{y_f-\delta_y^U}\int_{x_o+\delta_x^L}^{x_f-\delta_x^U}\left[\int_c^d\frac{(x-a_o)^i}{i!}\frac{(x-v)^{q-1}}{(q-1)!}\psi(c_o,v,y)f^{(i,q)}(a_o,v)\,dv\right]dx\,dy$$

$$-\sum_{i=0}^{p-1}\sum_k A_k\left[\int_c^d\frac{(x_k-a_o)^i}{i!}\frac{(y_k-v)^{q-1}}{(q-1)!}\psi(c_o,v,y)f^{(i,q)}(a_o,v)\,dv\right]_k$$

$$+\int_{y_o+\delta_y^L}^{y_f-\delta_y^U}\int_{x_o+\delta_x^L}^{x_f-\delta_x^U}\left[\int_a^b\int_c^d\frac{(x-u)^{p-1}}{(p-1)!}\psi(a_o,u,x)\frac{(y-v)^{q-1}}{(q-1)!}\psi(c_o,v,y)\right.$$

$$\left.\times f^{(p,q)}(u,v)\,du\,dv\right]dx\,dy$$

$$-\sum_k A_k\left[\int_a^b\int_c^d\frac{(x_k-u)^{p-1}}{(p-1)!}\psi(a_o,u,x)\frac{(y_k-v)^{q-1}}{(q-1)!}\psi(c_o,v,y)f^{(p,q)}(u,v)\,du\,dv\right]_k.$$

Here the limits of integration (a,b,c,d) correspond to the boundaries of the rectangle described in Definition 1. The order of the integration and summation can be rearranged to give:

$$\varepsilon(f)=\sum_{j=0}^{q-1}\int_a^b\left[\begin{array}{l}\displaystyle\int_{y_o+\delta_y^L}^{y_f-\delta_y^U}\int_{x_o+\delta_x^L}^{x_f-\delta_x^U}\frac{(y-c_o)^j}{j!}\psi(a_o,u,x)\frac{(x-u)^{p-1}}{(p-1)!}\,dx\,dy\\ \displaystyle-\sum_k A_k\frac{(y_k-c_o)^j}{j!}\psi(a_o,u,x)\frac{(x_k-u)^{p-1}}{(p-1)!}\end{array}\right]f^{(p,j)}(u,c_o)\,du$$

$$+\sum_{i=0}^{p-1}\int_{c}^{d}\left[\int_{y_o+\delta_y^L}^{y_f-\delta_y^U}\int_{x_o+\delta_x^L}^{x_f-\delta_x^U}\frac{(x-a_o)^i}{i!}\frac{(y-v)^{q-1}}{(q-1)!}\psi(c_o,v,y)\,dx\,dy - \sum_k A_k\frac{(x_k-a_o)^i}{i!}\frac{(y_k-v)^{q-1}}{(q-1)!}\psi(c_o,v,y)\right]f^{(i,q)}(a_o,v)\,dv$$

$$+\int_{a}^{b}\int_{c}^{d}\left[\int_{y_o+\delta_y^L}^{y_f-\delta_y^U}\int_{x_o+\delta_x^L}^{x_f-\delta_x^U}\frac{(x-u)^{p-1}}{(p-1)!}\psi(a_o,u,x)\frac{(y-v)^{q-1}}{(q-1)!}\psi(c_o,v,y)\,dx\,dy - \sum_k A_k\frac{(x_k-u)^{p-1}}{(p-1)!}\psi(a_o,u,x)\frac{(y_k-v)^{q-1}}{(q-1)!}\psi(c_o,v,y)\right]$$

$$\times f^{(p,q)}(u,v)\,du\,dv$$

or

$$\varepsilon(f)=\sum_{j=0}^{q-1}\int_{a}^{b}K_{p,j}(u,a_o,c_o)f^{(p,j)}(u,c_o)\,du + \sum_{i=0}^{p-1}\int_{c}^{d}K_{i,q}(v,a_o,c_o)f^{(i,q)}(a_o,v)\,dv + \int_{a}^{b}\int_{c}^{d}K_{p,q}(u,v,a_o,c_o)f^{(p,q)}(u,v)\,du\,dv\,,$$

where

$$K_{p,j}(u,a_o,c_o)=\int_{y_o+\delta_y^L}^{y_f-\delta_y^U}\int_{x_o+\delta_x^L}^{x_f-\delta_x^U}\frac{(y-c_o)^j}{j!}\psi(a_o,u,x)\frac{(x-u)^{p-1}}{(p-1)!}\,dx\,dy - \sum_k A_k\frac{(y_k-c_o)^j}{j!}\psi(a_o,u,x)\frac{(x_k-u)^{p-1}}{(p-1)!}$$

$$K_{i,q}(v,a_o,c_o)=\int_{y_o+\delta_y^L}^{y_f-\delta_y^U}\int_{x_o+\delta_x^L}^{x_f-\delta_x^U}\frac{(x-a_o)^i}{i!}\frac{(y-v)^{q-1}}{(q-1)!}\psi(c_o,v,y)\,dx\,dy - \sum_k A_k\frac{(x_k-a_o)^i}{i!}\frac{(y_k-v)^{q-1}}{(q-1)!}\psi(c_o,v,y)$$

$$K_{p,q}(u,v,a_o,c_o) = \int\limits_{y_o+\delta_y^L}^{y_f-\delta_y^U} \int\limits_{x_o+\delta_x^L}^{x_f-\delta_x^U} \frac{(x-u)^{p-1}}{(p-1)!}\psi(a_o,u,x)\frac{(y-v)^{q-1}}{(q-1)!}\psi(c_o,v,y)\,dx\,dy$$
$$- \sum_k A_k \frac{(x_k-u)^{p-1}}{(p-1)!}\psi(a_o,u,x)\frac{(y_k-v)^{q-1}}{(q-1)!}\psi(c_o,v,y)\,. \quad \text{(C.18)}$$

From Equation (C.16), we can write Equation (C.18) as:

$$K_{p,j}(u,a_o,c_o) = \varepsilon\left[\frac{(x-u)^{p-1}}{(p-1)!}\psi(a_o,u,x)\frac{(y-c_o)^j}{j!}\right] \qquad 0 \le j < p$$
$$K_{i,q}(v,a_o,c_o) = \varepsilon\left[\frac{(x-a_o)^i}{i!}\frac{(y-v)^{q-1}}{(q-1)!}\psi(c_o,v,y)\right] \qquad 0 \le i < q$$
$$K_{p,q}(u,v,a_o,c_o) = \varepsilon\left[\frac{(x-u)^{p-1}}{(p-1)!}\psi(a_o,u,x)\frac{(y-v)^{q-1}}{(q-1)!}\psi(c_o,v,y)\right]\,.$$

Theorem 3 is proven.

Theorem 4 (Error Estimates in an Integration Formula in R_2). *The error estimate of an integration formula in R_2 is*

$$|\varepsilon(f)| \le \sum_{j=0}^{q-1} e_{p,j}M_{p,j}(c_o) + \sum_{i=0}^{p-1} e_{i,q}M_{i,q}(a_o) + e_{p,q}M_{p,q}\,,$$

where

$$e_{p,j} = \int_a^b \left|K_{p,j}(u,a_o,c_o)\right|\,du$$
$$e_{i,q} = \int_c^d \left|K_{i,q}(v,a_o,c_o)\right|\,dv$$
$$e_{p,q} = \int_a^b\int_c^d \left|K_{p,q}(u,v,a_o,c_o)\right|\,dv\,du \qquad \text{(C.19)}$$

$$M_{p,j}(c_o) \equiv \sup_{u\in[a,b]} \left|f^{(p,j)}(u,c_o)\right|$$
$$M_{i,q}(a_o) \equiv \sup_{v\in[c,d]} \left|f^{(i,q)}(a_o,v)\right|$$
$$M_{p,q} \equiv \sup_{R_2} \left|f^{(p,q)}(u,v)\right|\,. \qquad \text{(C.20)}$$

Proof. From Equation (C.9), we can write:

$$\varepsilon(f) = \sum_{j=0}^{q-1} \int_a^b K_{p,j}(u, a_o, c_o) f^{(p,j)}(u, c_o)\, du + \sum_{i=0}^{p-1} \int_a^b K_{i,q}(v, a_o, c_o) f^{(i,q)}(a_o, v)\, dv + \int_a^b \int_c^d K_{p,q}(u, v, a_o, c_o) f^{(p,q)}(u, v)\, du\, dv$$

$$\left|\varepsilon(f)\right| \le \left| \sum_{j=0}^{q-1} \int_a^b K_{p,j}(u, a_o, c_o) f^{(p,j)}(u, c_o)\, du + \sum_{i=0}^{p-1} \int_a^b K_{i,q}(v, a_o, c_o) f^{(i,q)}(a_o, v)\, dv + \int_a^b \int_c^d K_{p,q}(u, v, a_o, c_o) f^{(p,q)}(u, v)\, du\, dv \right|$$

$$\le \left| \sum_{j=0}^{q-1} \int_a^b K_{p,j}(u, a_o, c_o) f^{(p,j)}(u, c_o)\, du \right| + \left| \sum_{i=0}^{p-1} \int_a^b K_{i,q}(v, a_o, c_o) f^{(i,q)}(a_o, v)\, dv \right| + \left| \int_a^b \int_c^d K_{p,q}(u, v, a_o, c_o) f^{(p,q)}(u, v)\, du\, dv \right|$$

$$\le \sum_{j=0}^{q-1} \int_a^b \left| K_{p,j}(u, a_o, c_o) f^{(p,j)}(u, c_o) \right| du + \sum_{i=0}^{p-1} \int_a^b \left| K_{i,q}(v, a_o, c_o) f^{(i,q)}(a_o, v) \right| dv + \int_a^b \int_c^d \left| K_{p,q}(u, v, a_o, c_o) f^{(p,q)}(u, v) \right| du\, dv$$

$$\le \sum_{j=0}^{q-1} \int_a^b \left| K_{p,j}(u, a_o, c_o) \right| \left| f^{(p,j)}(u, c_o) \right| du + \sum_{i=0}^{p-1} \int_a^b \left| K_{i,q}(v, a_o, c_o) \right| \left| f^{(i,q)}(a_o, v) \right| dv + \int_a^b \int_c^d \left| K_{p,q}(u, v, a_o, c_o) \right| \left| f^{(p,q)}(u, v) \right| du\, dv$$

$$\le \sum_{j=0}^{q-1} \int_a^b \left| K_{p,j}(u, a_o, c_o) \right| \sup_{u \in [a,b]} \left| f^{(p,j)}(u, c_o) \right| du + \sum_{i=0}^{p-1} \int_a^b \left| K_{i,q}(v, a_o, c_o) \right| \sup_{v \in [c,d]} \left| f^{(i,q)}(a_o, v) \right| dv + \int_a^b \int_c^d \left| K_{p,q}(u, v, a_o, c_o) \right| \sup_{R_2} \left| f^{(p,q)}(u, v) \right| du\, dv\,.$$

Defining

$$e_{p,j} = \int_a^b \left|K_{p,j}(u, a_o, c_o)\right| du \qquad M_{p,j}(c_o) \equiv \sup_{u\in[a,b]} \left|f^{(p,j)}(u, c_o)\right|$$

$$e_{i,q} = \int_c^d \left|K_{i,q}(v, a_o, c_o)\right| dv \qquad M_{i,q}(a_o) \equiv \sup_{v\in[c,d]} \left|f^{(i,q)}(a_o, v)\right|$$

$$e_{p,q} = \int_a^b \int_c^d \left|K_{p,q}(u, v, a_o, c_o)\right| dv\, du \quad M_{p,q} \equiv \sup_{R_2} \left|f^{(p,q)}(u, v)\right| ,$$

we obtain:

$$\left|\varepsilon(f)\right| \leq \sum_{j=0}^{q-1} e_{p,j} M_{p,j}(c_o) + \sum_{i=0}^{p-1} e_{i,q} M_{i,q}(a_o) + e_{p,q} M_{p,q} .$$

Theorem 4 is proven.

Corollary 1 (Error Estimates in the 2-D Noncollocated Simpson Rule for an Illuminance Function).
If $f(x,y) = E(x,y) = G(x,y) * H(x,y)$, *where* $E(x,y)$ *is the illuminance function,* $G(x,y) = \frac{1}{2\pi\sigma_b^2} \mathrm{e}^{-\frac{x^2+y^2}{2\sigma_b^2}}$ *is a 2-D Gaussian function representing the diffraction phenomenon, and* $H(x,y) \in \{0, \ldots, 1\}$ *is the optical flux function, then the error estimate of the 2-D noncollocated Simpson rule is*

$$\left|\varepsilon(f)\right| \leq h_x h_y \left(\sum_{j=0}^{2} e^N_{3,j} M_{3,j} + \sum_{i=0}^{2} e^N_{i,3} M_{i,3} + e^N_{3,3} M_{3,3} \right) , \tag{C.21}$$

where

$$e^N_{3,j} = e_{3,j} \frac{1}{h_x h_y}$$

$$e^N_{i,3} = e_{i,3} \frac{1}{h_x h_y}$$

$$e^N_{3,3} = e_{3,3} \frac{1}{h_x h_y}$$

are normalized error coefficients (Stroud 1971) with $e_{3,j}, e_{i,3}, e_{3,3}$ *and* $M_{3,j}, M_{i,3}, M_{3,3}$ *defined as in Equations* (C.19) *and* (C.20)*, respectively. Furthermore, the values of these coefficients are:*

$$e^N_{3,0} = e^N_{0,3} = 55.5 \times 10^{-3}$$

$$e^N_{3,1} = e^N_{1,3} = 0$$

$$e^N_{3,2} = e^N_{2,3} = 9.25 \times 10^{-3}$$

$$e^N_{3,3} = 5.4167 \times 10^{-3}$$

and

$$M_{3,0} = M_{0,3} = \frac{1.380}{s_b^3}$$
$$M_{3,1} = M_{1,3} = \frac{0.837}{s_b^4}$$
$$M_{3,2} = M_{2,3} = \frac{1.380}{s_b^5}$$
$$M_{3,3} = \frac{1.9047}{s_b^6} .$$

Proof. Given the results of Theorem 1 and Theorem 3, it can be easily concluded that for the noncollocated Simpson rule $p = q = 3$.

The normalized error coefficients $e^N_{3,j}, e^N_{i,3}, e^N_{3,3}$ are calculated using Theorem 4 and are tabulated in (Stroud 1971).

The derivation of the coefficients $M_{3,j}, M_{i,3}, M_{3,3}$ follows:

From the definition of an illuminance function we have:

$$E(x,y) = G(x,y) * H(x,y) . \tag{C.22}$$

Substituting Equation (C.22) into Equation (C.20), we obtain:

$$\begin{aligned} M_{3,j} &\equiv \sup_{u\in[a,b]} \left|E^{(3,j)}(u,c_o)\right| \qquad 0 \le j < 3 \\ &= \sup_{u\in[a,b]} \left|\frac{\partial^{3+j}}{\partial u^3 \partial v^j}\left(G(x,y) * H(x,y)\big|_{u=x,y=c_o}\right)\right| \\ M_{i,3} &\equiv \sup_{u\in[a,b]} \left|E^{(i,3)}(a_o,v)\right| \qquad 0 \le i < 3 \\ M_{3,3} &\equiv \sup_{u\in[a,b]} \left|E^{(3,3)}(u,v)\right| . \end{aligned} \tag{C.23}$$

The optical flux function that maximizes the convolution operation in Equation (C.23) is $H(x,y) = 1$:

$$\begin{aligned} M_{3,j} &= \sup_{u\in[a,b]} \left|\frac{\partial^{3+j}}{\partial u^3 \partial v^j} G(u,v)\big|_{y=c_o}\right| \qquad 0 \le j < 3 \\ M_{i,3} &= \sup_{u\in[a,b]} \left|\frac{\partial^{3+i}}{\partial u^i \partial v^3} G(u,v)\big|_{x=a_o}\right| \qquad 0 \le i < 3 \\ M_{3,3} &= \sup_{u\in[a,b]} \left|\frac{\partial^{3+3}}{\partial u^3 \partial v^3} G(u,v)\right| . \end{aligned} \tag{C.24}$$

Calculating the supremum of the multiple Gaussian derivatives and substituting in Equation (C.24), we get:

$$M_{3,0} = M_{0,3} = \frac{1.380}{s_b^3}$$
$$M_{3,1} = M_{1,3} = \frac{0.837}{s_b^4}$$
$$M_{3,2} = M_{2,3} = \frac{1.38}{s_b^5}$$
$$M_{3,3} = \frac{1.9047}{s_b^6} .$$

The corollary is proven.

The corollary makes it easy to design the subsampling grid size, defined by h_x and h_y. For a given desired maximum error $|e(f)|_{\max} \in 0, \ldots, 1$ and a smoothing level s_b determined by the camera system, using Equation (C.21) we can write:

$$h_x h_y = \frac{|\varepsilon(f)|_{\max}}{\left(\sum_{j=0}^{2} e_{3,j}^N M_{3,j} + \sum_{i=0}^{2} e_{i,3}^N M_{i,3} + e_{3,3}^N M_{3,3}\right)} . \tag{C.25}$$

For a squared subsampling grid, Equation (C.25) becomes:

$$h_x = h_y = \sqrt{\frac{|\varepsilon(f)|_{\max}}{\left(\sum_{j=0}^{2} e_{3,j}^N M_{3,j} + \sum_{i=0}^{2} e_{i,3}^N M_{i,3} + e_{3,3}^N M_{3,3}\right)}} . \tag{C.26}$$

Color Section

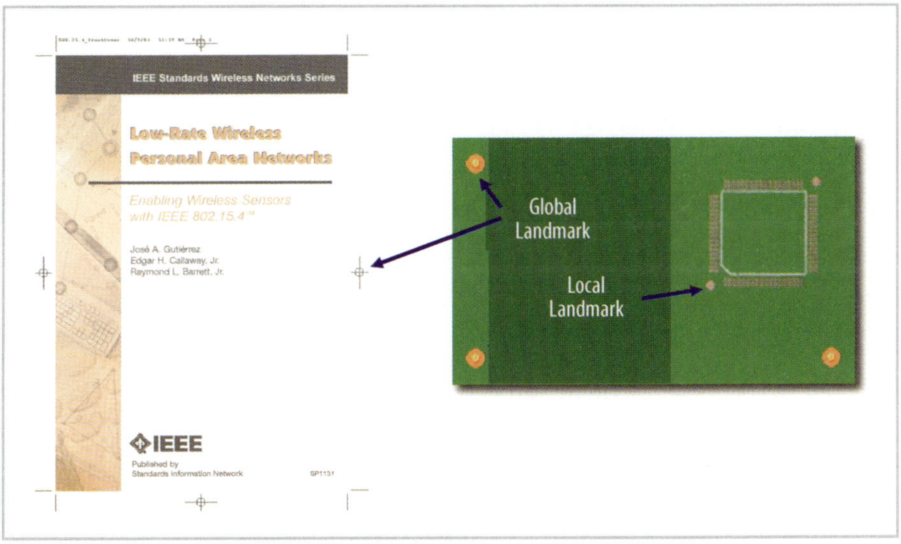

Plate 1. Landmarks use in the publishing and printed circuit board industries

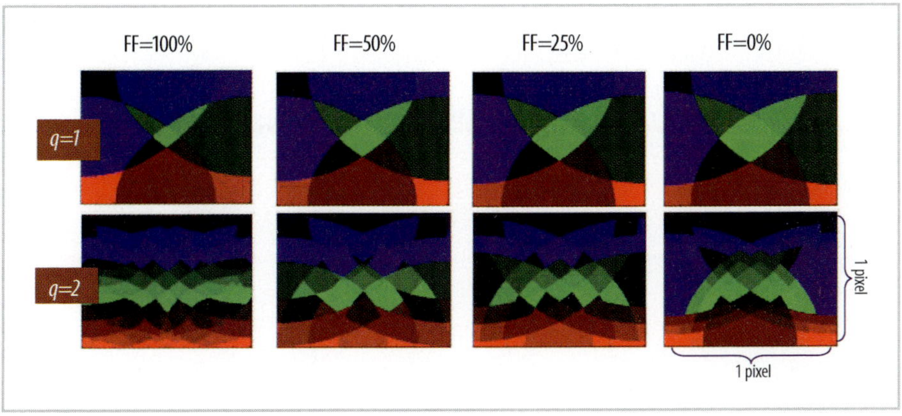

Plate 2. Example locale maps for four values of fill factor and to levels of intensity quantization

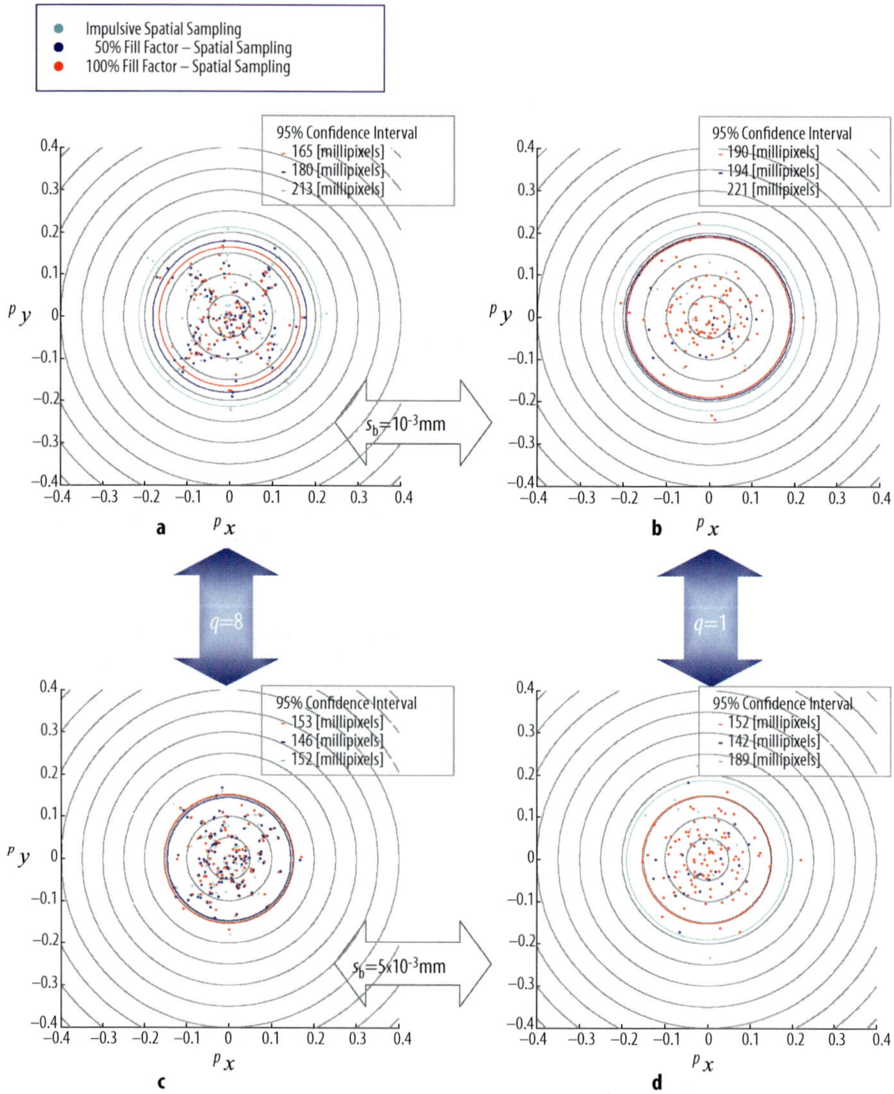

Plate 3. Effects of sensitive area, diffraction smoothing, and quantization

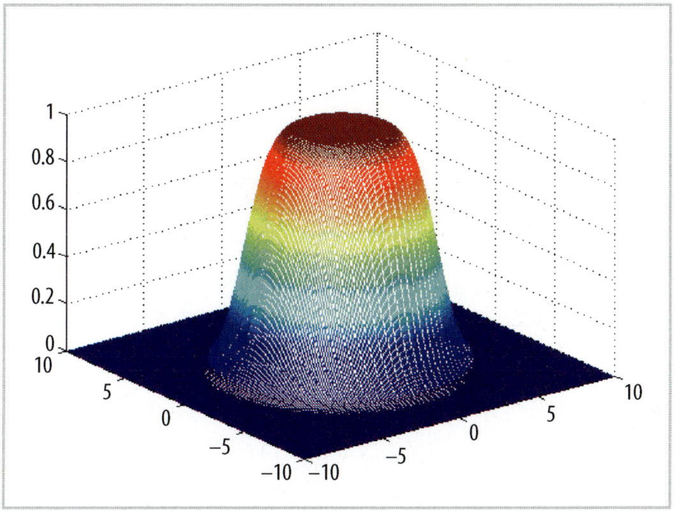

Plate 4. Illuminance function of a circular landmark

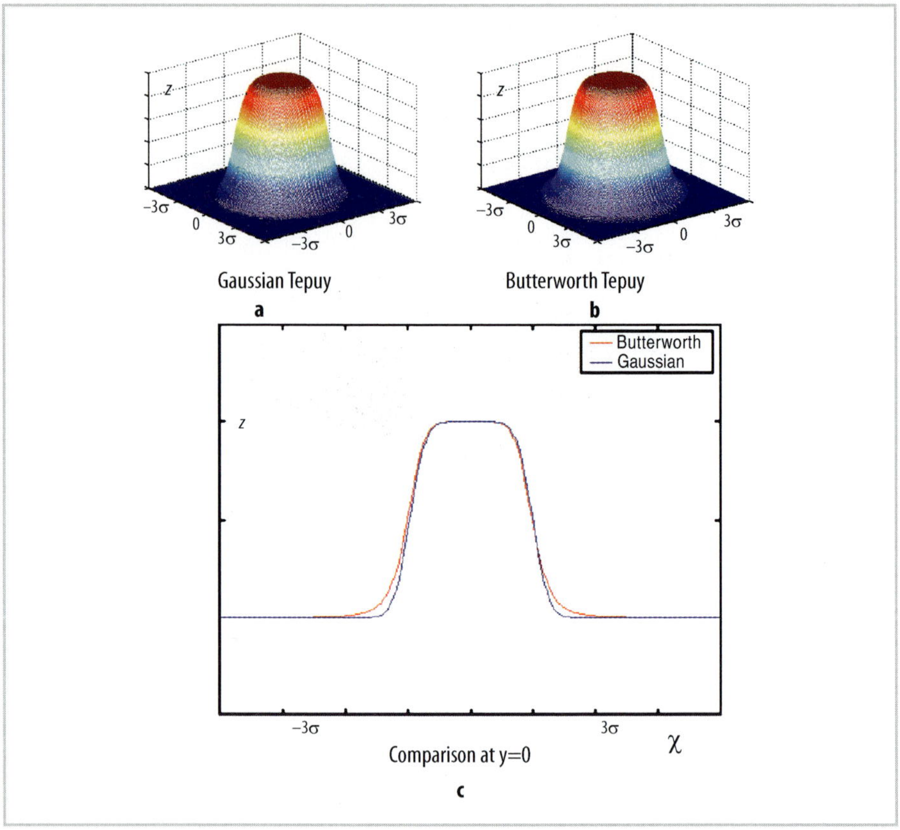

Plate 5. Comparison between a Gaussian Tepuy and a Butterworth Tepuy

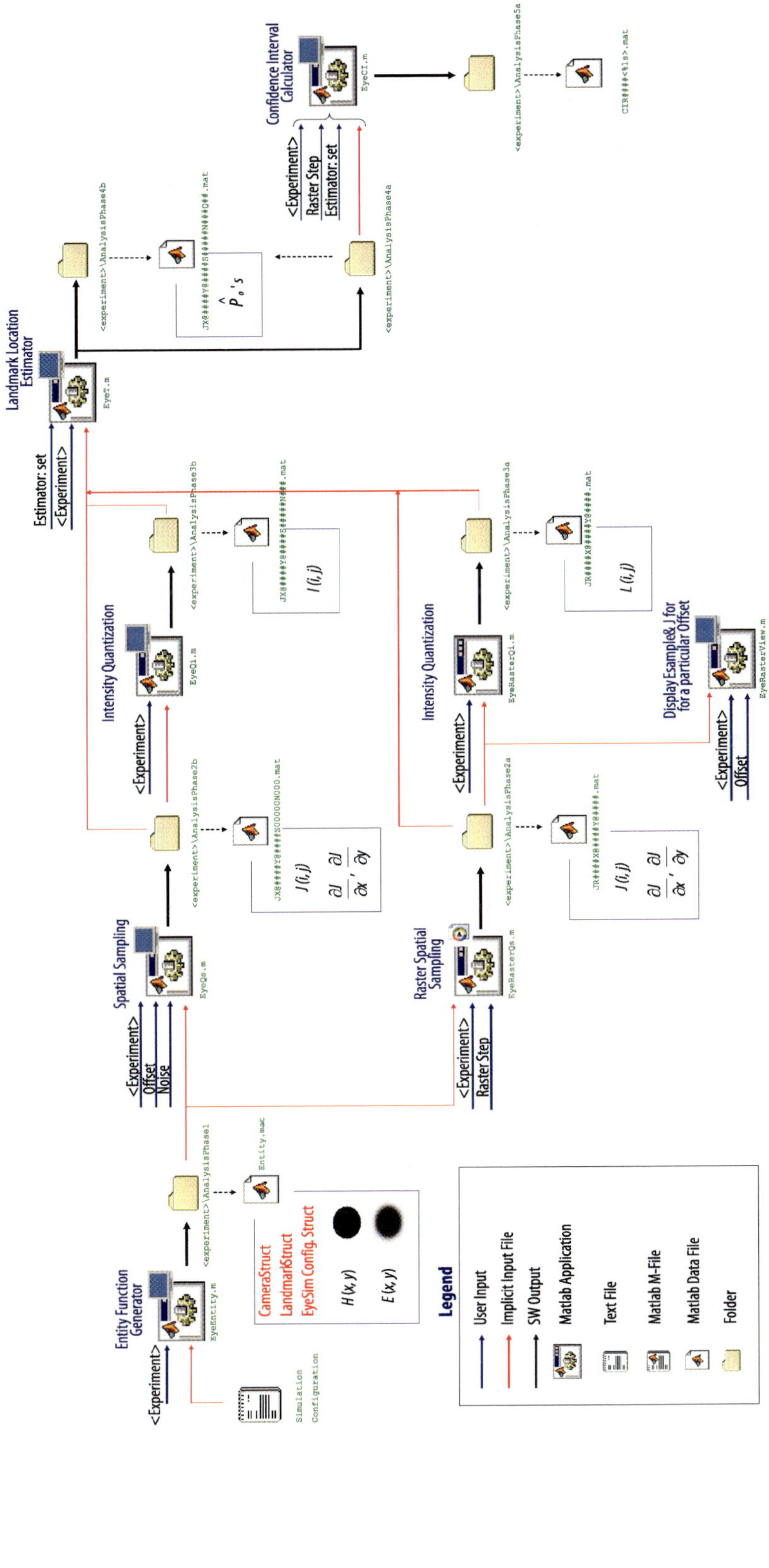

Plate 6. Functional architecture of the EyeBox Tool

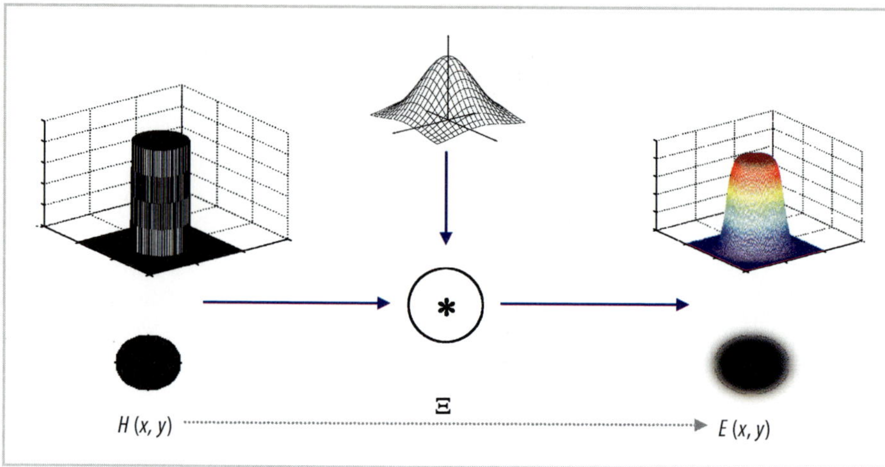

Plate 7. Convolution of the optical flux function with a Gaussian Kernel to produce the illummance function

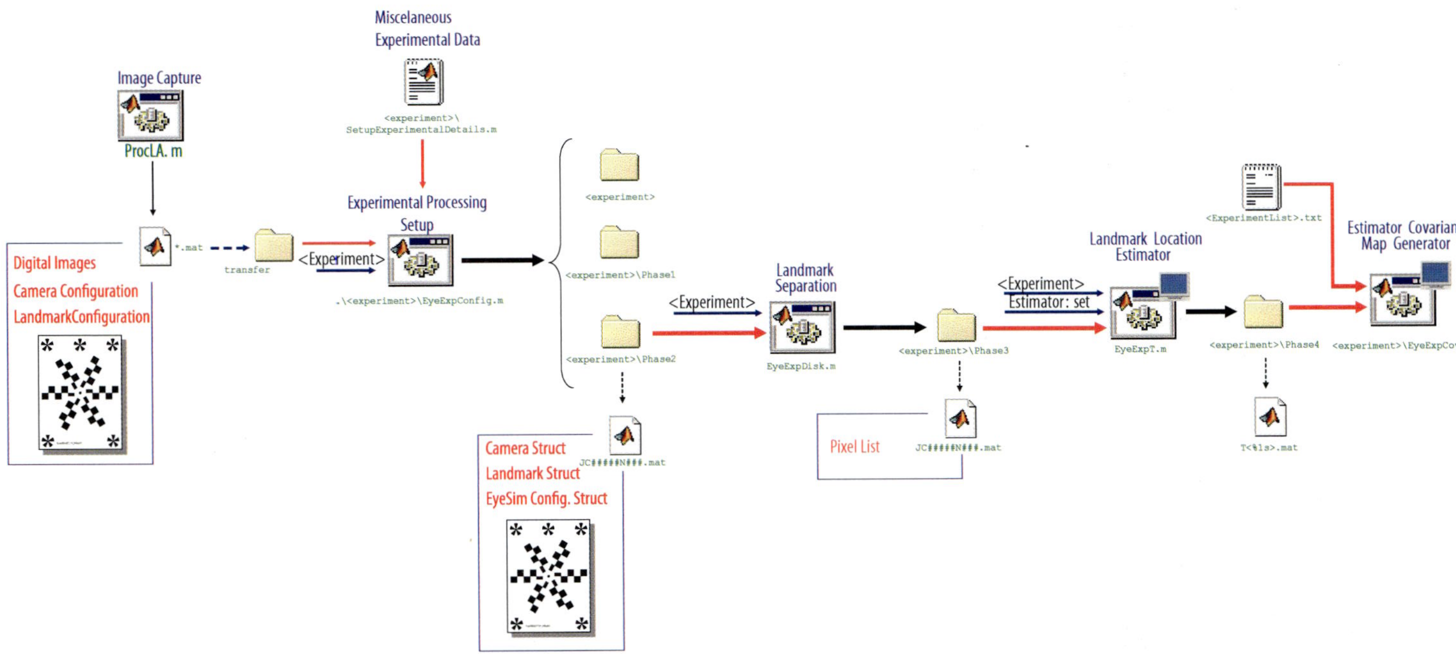

Plate 8. Functional architecture of the EyeExpBox tool

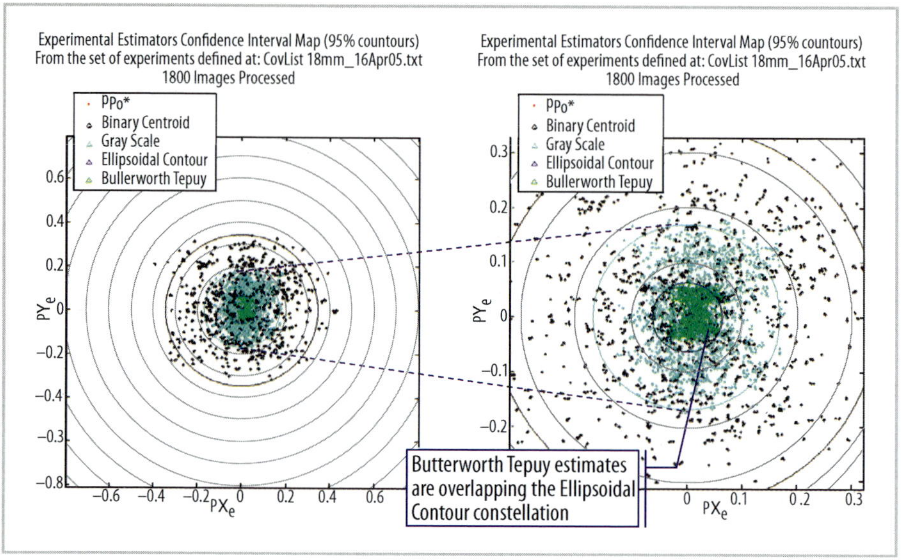

Plate 9. 18mm Configuration – Confidence Intervals – Constellation Details

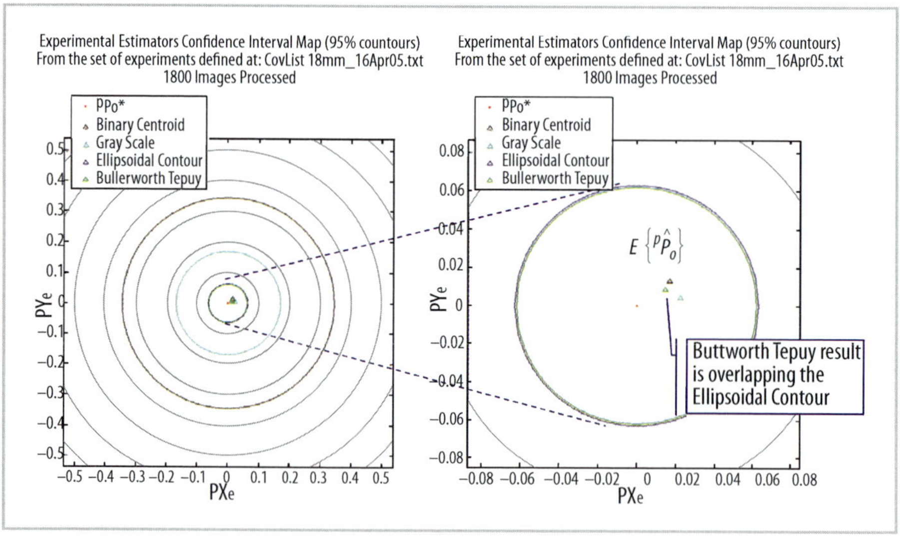

Plate 10. 18mm Configuration – Confidence Intervals

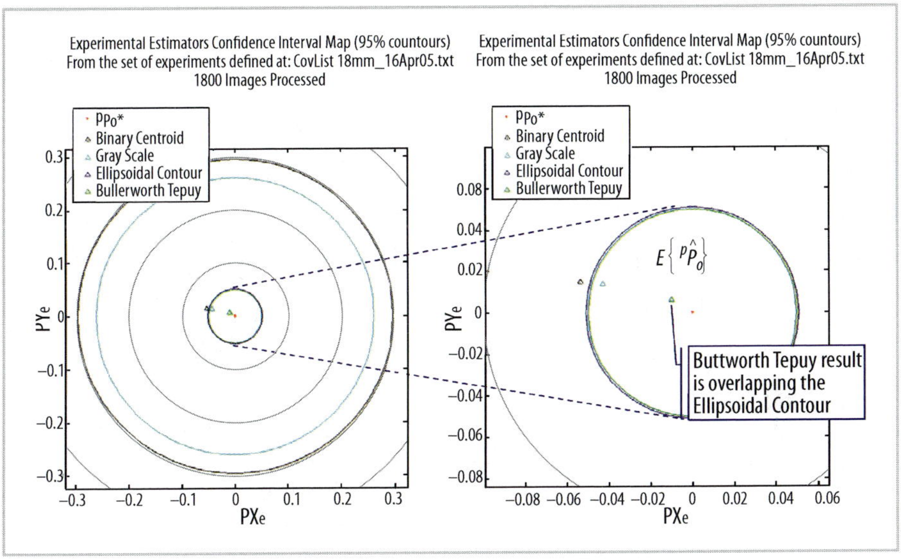

Plate 11. 35mm Configuration – Confidence Intervals

References

I. Amir, "An algorithm for finding the center of circular fiducials," Computer Vision, Graphics, and Image Processing, Vol. 49, No. 1, pp. 398–406, 1990.

B. Armstrong, B.S. Holeman, "Target tracking with a network of Doppler radars," IEEE Transactions on Aerospace and Electronic Systems, Vol. 34, No. 1, pp. 33–48, 1998.

B. Armstrong, T. Verron, L. Heppe, J. Reynolds, K. Schmidt, "RGR-3D: Simple, cheap detections of 6-DOF pose for tele-operation, and robot programming and calibration." In Proc. 2002 Int. Conf. on Robotics and Automation. IEEE: Washington, 2002.

K.B. Atkinson, "Close range photogrammetry and machine vision", Caithness: Whittles Publishing, 1996.

S. Baker, S. Nayar, H. Murase, "Parametric feature detection," Proceedings of IEEE Conference on CVPR '96, San Francisco, CA, 16–20 June 1996, p. 471.

D.H. Ballard, "Generalizing the Hough transform to detect arbitrary shapes," Pattern Recognition, Vol. 13, No. 2, pp. 111–122, 1981.

C.B. Bose, I. Amir, "Design of fiducials for accurate registration using machine vision," IEEE Transactions on Pattern Analysis and Machine Intelligence, Vol. 12, No. 12, pp. 1196–1200, 1990.

C.A. Berenstein, L.N. Kanal, D. Lavine, E.C. Olson, "A geometric approach to subpixel registration accuracy," Computer Vision, Graphics, and Image Processing, Vol. 40, No. 3, pp. 334–336, 1987.

A.M. Bruckstein, "On optimal image digitalization," IEEE Transactions on Acoustics, Speech, and Signal Processing, Vol. 35, No. 4, pp. 553–555, 1987.

A.M. Bruckstein, L. O'Gorman, A. Orlitsky, "Design of shapes for precise image registration," IEEE Transactions on Information Theory, Vol. 44, No. 7, pp. 3156–3162, 1998.

A.M. Bruckstein, R.J. Holt, "Optimum fiducials under weak perspective projections," Proceedings of the 7th IEEE International Conference on Computer Vision, Kerkyra, Vol. 1, No. 1, 20–27 September 1999.

G. Chiorboli, G.P. Vecchi, Comments on "Design of fiducials for accurate registration using machine vision," IEEE Transactions on Pattern Analysis and Machine Intelligence, Vol. 15, No. 12, pp. 1330–1332, 1993.

I.J. Cox, J.B. Kruskal, D.A. Wallach, "Predicting and estimating the accuracy of a subpixel registration algorithm," IEEE Transactions on Pattern Analysis and Machine Intelligence, Vol. 12, No. 8, pp. 721–734, 1990.

J.J. Craig, "Introduction to robotics," Reading, MA: Addison–Wesley, 1986.

M. Doros, "On the properties of the generation of discrete circular arcs on a square grid," Computer Vision, Graphics, and Image Processing, Vol. 28, No. 1, pp. 377–383, 1984.

L. Dorst, A.W.M. Smeulders, "Best linear unbiased estimators for properties of digitized straight lines," IEEE Transactions on Pattern Analysis and Machine Intelligence, Vol. 8, No. 2, pp. 279–282, 1986.

R.O. Duda, P.E. Hart, "Use of Hough transformation to detect lines and curves in pictures," Communications of the ACM, Vol. 15, No. 1, pp. 11–15, 1972.

V.N. Dvornychenko, "Bound on (deterministic) correlation functions with application to registration," IEEE Transactions on Pattern Analysis and Machine Intelligence, Vol. 5, No. 2, pp. 206–213, 1983.

A. Efrat, C. Gotsman, "Subpixel image registration using circular fiducials," International Journal on Computational Geometry and Applications, Vol. 4, No 4, pp. 403–422, 1994.

W. Förstner, "On the geometric precision of digital correlation," Proceedings of the International Archives of Photogrammetry and Remote Sensing (ISPRS) Commun. III Symp., Helsinki, Vol. 24, No. 3, pp. 176–189, 1982.

J. Goodman, "Introduction to Fourier optics," New York: McGraw–Hill Inc., 1968.

R.A. Gonsalves, "Cramér–Rao bounds on mensuration errors," Applied Optics, Vol. 15, No. 5, pp. 1270–1275, 1976.

R.C. Gonzalez, R.E. Woods, "Digital image processing," Boston: Addison–Wesley Publishing Company, 1992.

A. Hachicha, S. Simon, K. Hanna, "The use of gray-level information and fitting techniques for precise measurement of corneal curvature and thickness," Computer Vision and Pattern Recognition, Vol. 47, No. 1, pp. 131–164, 1989.

J. Hair, R. Anderson, R. Tatham, W. Black, "Multivariate data analysis with readings," 3rd edn., New York: Macmillan Publishing Company, 1992.

D.I. Havelock, "Geometric precision in noise-free digital images," IEEE Transactions on Pattern Analysis and Machine Intelligence, Vol. 11, No. 10, pp. 1065–1075, 1989.

D.I. Havelock, "The topology of locales and its effects on position uncertainty," IEEE Transactions on Pattern Analysis and Machine Intelligence, Vol. 13, No. 4, pp. 380–386, 1991.

M. Heath, S. Sarkar, T. Sanocki, K. Bowyer, "Comparison of edge detectors: a methodology and initial study," Proceedings of the 1996 IEEE Conference CVPR '96, San Francisco, CA, 16–20 June 1996, pp. 143–148.

E. Hecht, A. Zajac, "Optics," Boston: Addison–Wesley Publishing Company, 1979.

J.W. Hill, "Dimensional measurements from quantized images," in: D. Nitzan *et al.*, "Machine intelligence research applied to industrial automation," SRI 10th Report for NSF Grant DAR78-27128, Washington, DC: National Science Foundation, pp. 75–105, 1980.

C.-S. Ho, "Precision of digital vision systems," IEEE Transactions on Pattern Analysis and Machine Intelligence, Vol. 5, No. 6, pp. 593–601, 1983.

P.D. Hyde, L.S. Davis, "Subpixel edge estimation," Pattern Recognition, Vol. 16, No. 4, pp. 413–420, 1983.

H. Hung, "Lower bound on average mean-square error for image restoration," IEEE Transactions on Signal Processing, Vol. 39, No. 2, pp. 497–499, 1991.

IEEE SA, "IEEE 100, The authoritative dictionary of IEEE standard terms," 7th edn, Los Alamitos, CA: IEEE Press, 2000.

B. Jahne, "Practical handbook on image processing for scientific applications," Boca Raton, FL: CRC Press, 1997.

S.M. Kay, "Fundamentals of statistical signal processing: estimation theory," Upper Saddle River, NJ: Prentice–Hall, 1993.

N. Kiryati, A. Bruckstein, "Gray-levels can improve the performance of binary image digitizers," Computer Vision and Pattern Recognition, Vol. 53, No. 1, pp. 31–39, 1991.

H. Klaasman, "Some aspects of the accuracy of the approximated position of a straight line on a square grid," Computer Vision, Graphics, and Image Processing, Vol. 4, No.1, pp. 225–235, 1975.

A. Krommer, C.W. Ueberhuber, "Computational integration," Philadelphia, PA: Society for Industrial and Applied Mathematics Press, 1998.

Z. Kulpa, "On the properties of discrete circles, rings, and disks," Computer Vision, Graphics, and Image Processing, Vol. 10, No. 1, 1979, pp. 348–365.

C.R. McCluer, "Industrial mathematics: modeling in industry, science, and government," Upper Saddle River, NJ: Prentice–Hall, 2000.

A. Nakamura, K. Aizawa, "Digital circles," Computer Vision, Graphics, and Image Processing, Vol. 26, No. 1, pp. 242–255, 1984.

V.S. Nalwa, T.O. Binford, "On detecting edges," IEEE Transactions on Pattern Analysis and Machine Intelligence, Vol. 8, No. 6, pp. 699–714, 1986.

L. Nielsen, K. Astrom, E. Jury, "Optimal digitization of 2-D images," Transactions on Acoustics, Speech, and Signal Processing, Vol. 32, No. 6, pp. 1247–1249, 1984.

H.K. Nishihara, P.A. Crossley, "Measuring photolithographic overlay accuracy and critical dimensions by correlating binarized Laplacian and Gaussian convolution," IEEE Transactions on Pattern Analysis and Machine Intelligence, Vol. 10, No. 1, pp. 17–30, 1988.

L. O'Gorman, A.M. Bruckstein, C.B. Bose, I. Amir, "Subpixel registration using concentric ring fiducial," Proceedings of the 10th International Conference on Pattern Recognition, Atlantic City, NJ, 16–21 June 1990, pp. 249–253.

L. O'Gorman, "Subpixel precision of straight-edged shapes for registration and measurement," IEEE Transactions on Pattern Analysis and Machine Intelligence, Vol. 18, No. 7, pp. 746–751, 1996.

P.Z. Peebles, "Probability, random variables, and random signal principles," 4th edn., New York: McGraw–Hill Inc., 2001.

J.G. Proakis, D.G. Manolakis, "Digital signal processing: principles, algorithms, and applications," 3rd edn., Upper Saddle River, NJ: Prentice–Hall, 1996.

A. Reza, M. Doroodchi, "Cramér–Rao lower bound on locations of sudden changes in a steplike signal," IEEE Transactions on Signal Processing, Vol. 44, No. 16, pp. 2551–2556, 1996.

D. Robinson, P. Milanfar, "Fundamental performance limits in image registration," IEEE Transactions on Image Processing, Vol. 13, No. 9, pp. 1185–1199, 2004.

K. Rohr, "Landmark-based image analysis: using geometric and intensity models," Amsterdam: Kluwer Academic Publishers, 2001.

T.W. Ryan, R.T. Gray, B.R. Hunt, "Prediction of correlation errors in stereopairs images," Optical Engineering, Vol. 19, No. 3, pp. 312–322, 1982.

L.L. Scharf, "Statistical signal processing, detection, estimation, and time series analysis," Boston, MA: Addison–Wesley Publishing Company, 1990.

L.L. Scharf, L.T. McWhorter, "Geometry of the Cramér–Rao bound," Signal Processing, Vol. 31, No. 3, pp. 1–11, 1993.

F.W. Sears, "Optica," Boston, MA: Addison–Wesley, 1979.

S.D. Shapiro, "Properties of transforms for the detection of curves in noisy pictures," Computer Graphics and Image Processing, Vol. 8, No. 1, pp. 219–236, 1978.

R. Sriraman, J. Koplowitz, S. Mohan, "Tree searched chain coding for subpixel reconstruction of planar curves," IEEE Transactions on Pattern Analysis and Machine Intelligence, Vol. 11, No. 1, pp. 95–104, 1989.

S. Stewart, "Calculus: early transcendentals," Pacific Grove, CA: Brooks/Cole Publishing Co., 1994.

A.H. Stroud, "Approximate calculation of multiple integrals," Upper Saddle River, NJ: Prentice-Hall, 1971.

A.J. Tabatabai, O.R. Mitchell, "Edge detection to subpixel values in digital imagery," IEEE Transactions on Pattern Analysis and Machine Intelligence, Vol. 6, No. 2, pp. 181–201, 1984.

J.D. Thurgood, M.E. Mikhail, "Photogrammetric analysis of digital images," International Archives of Photogrammetry, Vol. 24, No. 3, pp. 576–590, 1982.

Q. Tian, M.N. Huhns, "Algorithms for subpixel registration," Computer Vision, Graphics, and Image Processing, Vol. 35, No. 2, pp. 220–233, 1986.

M. Tichem, M.S. Cohen, "Sub-µm registration of fiducial marks using machine vision," IEEE Transactions on Pattern Analysis and Machine Intelligence, Vol. 16, No. 8, pp. 791–794, 1994.

H.L. Van Trees, "Detection, estimation, and modulation theory, Part 1," New York: John Wiley & Sons, Inc., 1968.

A.M. Vossepoel, A.W.M. Smeulders, "Vector code probability and metrication error in the representation of straight lines of finite length," Computer Vision, Graphics, and Image Processing, Vol. 20, No. 1, pp. 347–364, 1982.

B. Widrow, "A study of rough amplitude quantization by means of Nyquist sampling theory," IRE Transactions on Circuit Theory, Vol. 3, No. 1, pp. 266–276, 1965.

S. Yam, L.S. Davis, "Image registration using generalized Hough transforms," Proceedings of IEEE Computer Society on Pattern Recognition and Image Processing, Dallas, TX, 3–5 August 1981, pp. 526–533.

J. Zheng, M. Lei, "Method and apparatus for locating an acquisition target in two-dimensional images by detecting symmetry in two different directions," United States Patent 5,515,447, 1996.

J.Y. Zhou, D. Lopresti, P. Sarkar, G. Nagy, "Spatial sampling effects on scanned 2-D patterns," IEEE Transactions on Pattern Analysis and Machine Intelligence, Vol. 20, No. 3, 1998.

R.E. Ziemmer, W.H. Tranter, D.R. Fannin, "Signals and systems: continuous and discrete," New York: Macmillan Publishing Co., 1983.

Index

Printing: Krips bv, Meppel, The Netherlands
Binding: Stürtz, Würzburg, Germany